T0221103

Climate Dynamics

CLIMATE DYNAMICS

KERRY H. COOK

PRINCETON UNIVERSITY PRESS

PRINCETON AND OXFORD

Published by Princeton University Press, 41 William Street, Princeton, New Jersey 08540
In the United Kingdom: Princeton University Press, 6 Oxford Street, Woodstock,
Oxfordshire OX20 1TW

press.princeton.edu

Library of Congress Cataloging-in-Publication Data

Cook, Kerry Harrison, 1953–
Climate dynamics / Kerry H. Cook.
     p. cm.
Includes bibliographical references and index.
ISBN 978-0-691-12530-5 (hardcover : alk. paper) 1. Climatology. 2. Human ecology. I. Title.
QC861.3.C66 2013
551.6—dc23
     2012032005

British Library Cataloging-in-Publication Data is available

This book has been composed in Sabon and Trajan

Printed on acid-free paper. ∞

Printed in the United States of America

10 9 8 7 6 5 4 3 2 1

Dedicated to my children,
Hilary and Jeffrey,
And in memory of my mother, Alva

CONTENTS

Preface

The purpose of this book is to provide a foundation in the physical understanding of the earth's climate system. It is based on more than 20 years' experience teaching climate dynamics at Cornell University and, more recently, The University of Texas at Austin.

Most universities and colleges do not have a Department of Atmospheric Science, in which a technical course on climate dynamics would most likely be found. However, there is a pressing need to increase and promote an understanding of the climate system. Climate change will affect all fields of study and all professions in the coming decades, including the medical and engineering professions, and it will increasingly involve global political systems and economic planning. I wrote this book to support and broaden the teaching of climate dynamics.

The book assumes no background in atmospheric or ocean sciences and is written to be accessible for teaching by faculty in any field of science, mathematics, or engineering. The material is appropriate for any science or engineering undergraduate student who has completed two semesters of calculus and one semester of calculus-based physics. In combination with selected readings from the most recent Assessment Report of the Intergovernmental Panel on Climate Change, this book can be used to develop a course on contemporary climate change that emphasizes the physical understanding of the climate system.

The first section of the book (chapters 1–3) provides a description of the climate system based on observations of the mean climate state and its variability. It introduces the vocabulary of the field, the dependent variables that describe the climate system, and the typical approaches taken in displaying these variables. Taken together, chapters 2 and 3 form an atlas of the climate system, and figures from these chapters are referenced throughout the book.

The second section of the book (chapters 4–6) is aimed at developing a quantitative understanding of the processes that determine the climate state—radiation, heat balances, and the basics of geophysical fluid dynamics. The fluid dynamics application is developed on the premise that the student is familiar with Newton's laws of motion and basic thermodynamics from a first course in college physics. With an understanding of the basic processes, applications for the atmosphere, ocean, and hydrologic cycle are developed in chapters 7–9. The last three chapters of the book, chapters 10–12, are more directly related to contemporary climate change.

Many people contributed to this book, directly and indirectly. My daughter, Hilary, and my son, Jeffrey, always keep life interesting, and I appreciate their love and support. I also thank my father, Robert Harrison, for many years of encouragement. Professor Peter Gierasch taught the class with me for a number of years at Cornell, and his clear-sighted treatment of the material is reflected especially in the chapters on radiation and ocean dynamics. Major contributions to creating and improving figures came from Dr. Edward Vizy, Zachary Launer, and I especially thank Meredith Brown, who also helped with technical editing. I am also grateful to Ingrid Gnerlich and her colleagues at Princeton University Press for their patience and persistence. The hundreds of students in my climate classes over the years inspired this book and influenced its contents through their insightful questions, eagerness to *really* understand, and their genuine concern for the future of the planet. I am grateful to them all.

CLIMATE DYNAMICS

An Introduction to the Climate System

Climate dynamics is the scientific study of how and why climate changes. The intent is not to understand day-to-day changes in weather but to explain average conditions over many years. Climate processes are typically associated with multidecadal time scales, and continental to global space scales, but one can certainly refer to the climate of a particular city.

Climate dynamics is a rapidly developing field of study, motivated by the realization that human activity is changing climate. It is necessary to understand the natural, or unperturbed, climate system and the processes of human-induced change to be able to forecast climate so that individuals and governments can make informed decisions about energy use, agricultural practice, water resources, development, and environmental protection.

Climate has been defined as "the slowly varying aspects of the atmosphere/hydrosphere/lithosphere system."[1] Other definitions of climate might also explicitly include the biosphere as part of the climate system, since life on the planet plays a well-documented role in determining climate. Anthropogenic climate change is just one example, but there are others, such as the influence of life on the chemical composition of the atmosphere throughout its 4.5 billion–year life span.

The word *climate* is derived from the Greek word *klima*, which refers to the angle of incidence of the sun. This is a fitting origin because solar radiation is the ultimate energy source for the climate system. But to understand climate we need to consider much more than solar heating. Processes within the earth system convert incoming solar radiation to other forms of energy and redistribute it over the globe from pole to pole and throughout the vertical expanses of the atmosphere and ocean. This energy not only warms the atmosphere and oceans but also fuels winds and ocean currents, activates phase changes of water, drives chemical transformations, and supports biological activity. Many interacting processes create the variety of climates found on the earth.

A schematic overview of the global climate system is provided in Figure 1.1. This diagram represents the climate system as being composed of five subsystems—the atmosphere, the hydrosphere, the biosphere, the cryosphere, and the land surface. It also depicts processes that are important for determining the climate state, such as the exchange of heat, momentum, and water among the subsystems, and represents the agents of climate change.

---

[1] From the *Glossary of Meteorology*, published by the American Meteorological Society.

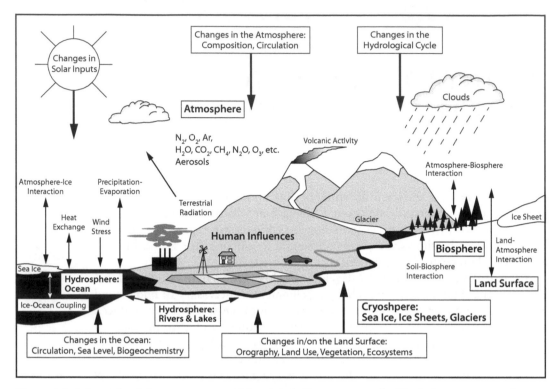

Figure 1.1 Schematic of the components of the global climate system (bold), their processes and interactions (thin arrows) and some aspects that may change (bold arrows). From IPCC, 2001.

Figure 1.1 provides an excellent summary of the climate system, and it is useful as a first-order, nontechnical description. At the other end of the spectrum is the *Bretherton diagram*, shown in Figure 1.2. This detailed, perhaps a bit overwhelming, schematic was constructed to characterize the full complexity of climate. It is a remarkable and rich representation of the system, illustrating the many processes that influence climate on all time scales. It coalesces historically separate fields of scientific inquiry—demonstrating that not only atmospheric science and oceanography are relevant to climate science but that various subdisciplines of geology, biology, physics, and chemistry—as well as the social sciences—are all integral to an understanding of climate.

This is a very exciting and critical time in the field of climate dynamics. There is reliable information that past climates were very different from today's climate, so we know the system is capable of significant change. We also understand that it is possible for the system to change quickly. The chemical composition of the atmosphere is changing before our eyes, and satellite- and earth-based observing networks allow us to monitor changes in climate fairly accurately.

Clearly, this one text on climate dynamics cannot cover the full breadth of this wide-ranging and rapidly developing field, but it provides the reader with the fundamentals—the background needed for a basic understanding of

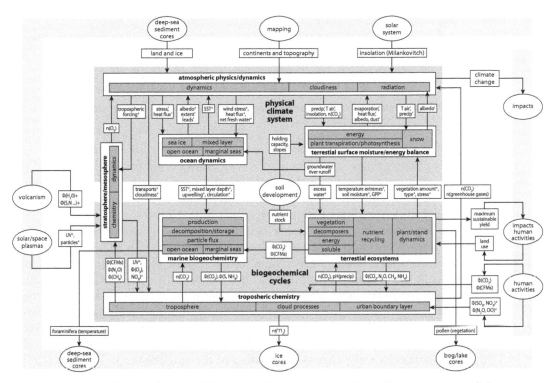

Figure 1.2 The Bretherton diagram, illustrating the components of the climate system and the interactions among them. (' – on timescale of hours to days; * = on timescale of months to seasons; ϕ = flux; n = concentration; SST is sea surface temperature)

climate and climate change, and a launchpad for reading the scientific literature and, it is hoped, contributing to the profound challenge before humanity of managing climate change. With this fundamental understanding, science can address the questions, needs, and constraints of society in a reasonable and useful way, and offer informed answers to guide society's behavior.

REFERENCE AND ADDITIONAL READING

IPCC, 2001: Climate Change 2001: The Scientific Basis. Report of the Intergovernmental Panel on Climate Change. Houghton, J. T., Y. Ding, D. J. Griggs, M. Noguer, P. J. van der Linden, X. Dai, K. Maskell, and C. A. Johnson (eds.). Cambridge University Press, Cambridge and New York.

2
THE OBSERVED CLIMATOLOGY

This chapter forms a concise atlas of the climate system. An overview of the system is presented using the variables and terminology commonly used to characterize climate. These terms are referenced in subsequent chapters as a deeper understanding of climate processes is developed.

Some features of the climate system are known accurately, while others are known only approximately. Climate observations can be limited by insufficient spatial and temporal resolution, inadequate global coverage, or a lack of long-term records. Precipitation observations are a good example. Because of the high variability of precipitation over a wide range of space and time scales, the observing requirements for establishing a precipitation climatology are demanding. Global measurements of precipitation or, more accurately, measurements of radiative fluxes that can be translated into rainfall rates have been available only since the beginning of the satellite era in the early 1970s. Pre-satellite coverage over vast regions of the oceans was particularly sparse, especially in regions where ships rarely traveled. Establishing a climatology for other variables, such as evaporation and soil moisture, is even more challenging.

Many of the figures in this text were drawn using *reanalysis* products, which combine simulations using state-of-the-art numerical models with observations. To produce a reanalysis climatology, computer models are run to stimulate many decades, with observed fields incorporated into the model at the time they were observed. This process is called *four-dimensional data assimilation*, for the three spatial dimensions plus time. (Data assimilation is also used routinely in generating weather forecasts.) Thus, the reanalysis product is not pure observations but a blend of observations and computer model output. Reanalysis values of variables that are assimilated—for example, winds and temperatures—are accurate. Other variables, however, are model-dependent output and may not be as reliable. Sometimes, as in the case of evaporation, the reanalysis product is the best information available with global coverage. For other variables, ground-based and satellite observations, if available, are preferred to the reanalysis product.

Maps of climate variables use latitude and longitude as coordinates with an equidistant cylindrical projection. Keep in mind that the area of middle and high latitudes is falsely large in this projection. In reality, half the surface area of the globe lies between 30°N and 30°S latitude, whereas in the figures this region occupies only one-third of the area. Vertical profiles of climate variables are also shown, averaged globally or over certain regions using area weighting to correctly account for the decreasing distance between meridians (lines of constant longitude) away from the equator. Another useful way to display climate variables is as the *zonal mean*, in which the variable is averaged over all longitudes and displayed in the latitude/height plane.

The international system of units (SI) is used as reviewed in Appendix A.

2.1 THE ATMOSPHERE

We begin our description of the atmosphere with air pressure. Pressure is defined as "force per unit area" and is expressed in SI units of pascals (abbreviated Pa). Pressure is simply the weight of the overlying mass, m, per unit area:

$$p \equiv \frac{mg}{\text{area}} \quad \Rightarrow \quad \text{Pa} \sim \frac{\text{newton}}{\text{m}^2} = \frac{(\text{kg} \cdot \text{m})/\text{s}^2}{\text{m}^2} = \frac{\text{kg}}{\text{m} \cdot \text{s}^2}, \qquad (2.1)$$

where g is the acceleration due to gravity. Figure 2.1 shows the global distribution of surface pressure in units of hectapascals (hPa), where $100 \text{ Pa} = 1 \text{ hPa}$. This figure is not helpful for learning about the atmosphere, however, because surface topography dominates the distribution. Surface pressure is lowest over the highest mountains, and high and uniform over the oceans, because the overlying air column is thinner (less massive) at higher elevations. Consequently, surface pressures in the Himalayan Mountains and over Antarctica drop below 600 hPa but are close to 1000 hPa everywhere over the world's oceans.

Figure 2.1 demonstrates the close connection between pressure and elevation. Pressure is often used as a vertical coordinate in describing the atmosphere, replacing elevation, z. The average relationship between p and z in the earth's atmosphere, typical of large space and time scales, is in Figure 2.2. Note that p is not a linear function of z, that is, $p \neq az + b$, where a and b are constants. Instead, pressure decreases exponentially with height.

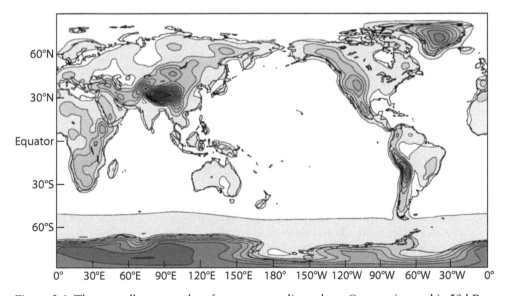

Figure 2.1 The annually averaged surface pressure climatology. Contour interval is 50 hPa.

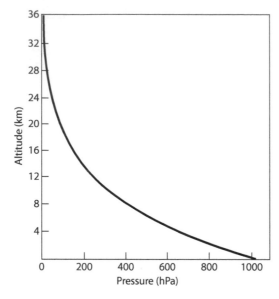

Figure 2.2 The average relationship between atmospheric pressure and altitude.

In the figures that follow, atmospheric variables are presented on surfaces of constant pressure, or *isobars*, instead of surfaces of uniform elevation. Figure 2.2 can be used to estimate the altitude of any pressure surface. Where topography extends up into the atmosphere, certain pressure levels may not exist. For example, since the surface pressure over Antarctica is 700 hPa or lower (see Fig. 2.1), there is no 900 hPa surface. Such regions may be specified as having missing data, or the data may be extrapolated to fill in the missing values.

When pressure is substituted for height as an *independent* variable, then the height of the pressure level becomes a *dependent* variable. (Recall that independent variables are the coordinate axes, and dependent variables describe the system. For the atmosphere, temperature and wind speeds are examples of dependent variables.) It is common to use *geopotential height*, Z, as the independent variable instead of height, z, where

$$Z \equiv \frac{1}{g_0} \int_0^z g \, dz. \tag{2.2}$$

In Eq. 2.2, g_0 is the acceleration due to gravity at the surface of the earth. Because the gravitational attraction between two bodies depends on the distance between them, the acceleration due to gravity decreases with increasing height—or decreasing pressure—in the atmosphere. At the earth's surface, $g = g_0 = 9.81$ m/s^2. At 10 km elevation, $g = 9.77$ m/s^2. This 0.4% reduction in g within the lower atmosphere is relatively small, so g can be taken as constant in Eq. 2.2. With this assumption, geopotential height, Z, can be interpreted as the elevation, z, of a pressure level.

Geopotential height is closely related to the gravitational potential energy, Φ, given by

$$\Phi(z) = \int_0^z g \, dz. \tag{2.3}$$

(Recall that work, which is a form of energy, is "force × distance".) Then Φ evaluated at some altitude z is the work that was done against gravity to lift a unit mass of air from the surface to that altitude. Equivalently, it is the "potential" energy that would be extracted if the unit mass were to fall to the surface. In common practice, Φ is referred to simply as the *geopotential*.

The annual mean geopotential height climatology at 900 and 200 hPa is displayed in Figure 2.3. Note the following:

- The 900 hPa surface is roughly 800 m from the surface in the subtropics (near 30°N and 30°S) and slopes down closer to the surface at higher latitudes and near the equator.

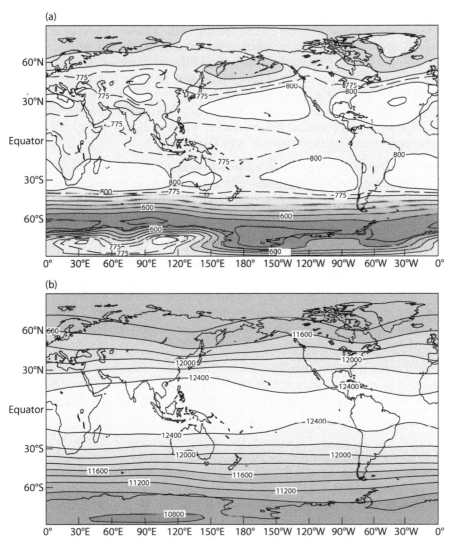

Figure 2.3 Annual mean geopotential height climatology in meters at 900 and 200 hPa. (a) The contour interval is 50 m; the 775 m contour is indicated with the dashed line. (b) The contour interval is 200 m.

- Geopotential height surfaces at 200 hPa are more than 12 km from the surface near the equator and slope down below 11 km at higher latitudes in both hemispheres.
- At 900 hPa, the highest geopotential heights are preferentially located over the oceans.
- The *equatorial trough* is the region of relatively low geopotential heights deep in the tropics.
- Poleward of the *subtropical highs* in both hemispheres, geopotential heights decrease through the midlatitudes all the way to the poles. (Geopotential height values at 900 hPa over East Antarctica are not realistic because the topography rises above this level.)
- The presence of the continents disturbs the zonal (east/west) uniformity of the geopotential height lines.

The distribution of geopotential heights has strong seasonal dependence, particularly at lower levels. To represent seasonal changes in this field and others, climatologies for the December, January, and February mean, designated DJF, are used to display Northern Hemisphere winter and Southern Hemisphere summer. For the opposite season, June, July, and August averages are displayed, and denoted as JJA. Seasonal ranges are quantified by plotting the difference DJF–JJA.

DJF and JJA geopotential heights at 900 hPa are shown in Figures 2.4a and b, respectively, and their difference is in Figure 2.4c.

- Geopotential heights are higher over the oceans than over the land in the hemisphere experiencing summer; the opposite is the case in the winter hemisphere.
- Seasonal differences in geopotential heights are greater over land surfaces than over the oceans.
- The subtropical highs centered over the oceans that were noted in the annual mean (Fig. 2.3) are primarily a summer feature in the Northern Hemisphere.
- In the Southern Hemisphere, geopotential heights are more zonally uniform in the winter (JJA), and located slightly closer to the equator.
- The monsoon regions of the world—southeastern Asia, northern Africa, tropical South America, the southwestern United States, and Australia—are characterized by low geopotential heights in the summer.

Geopotential height distributions at 200 hPa are displayed in Figure 2.5. The lines of constant geopotential height are much more zonally uniform than at the 900 hPa level, but some effects of continentality are still discernible.

- The low-level monsoon lows at 900 hPa are overlain by regions of high geopotential heights. This means that the distance between the 900 hPa and 200 hPa levels, known as the *thickness,* is greater in these regions.
- North of about 45°N in DJF there is a wavy pattern in the geopotential height lines, with two wave cycles encircling the globe. This pattern brings low heights to lower latitudes on the east coasts of Asia and North America, along with high geopotential height gradients. These are the *storm tracks* of the Northern Hemisphere, running southwest to northeast over the North Pacific and Atlantic Oceans.
- The most pronounced seasonal differences at this level occur over northeastern Asia (Siberia) and, to a lesser degree, northern North America.

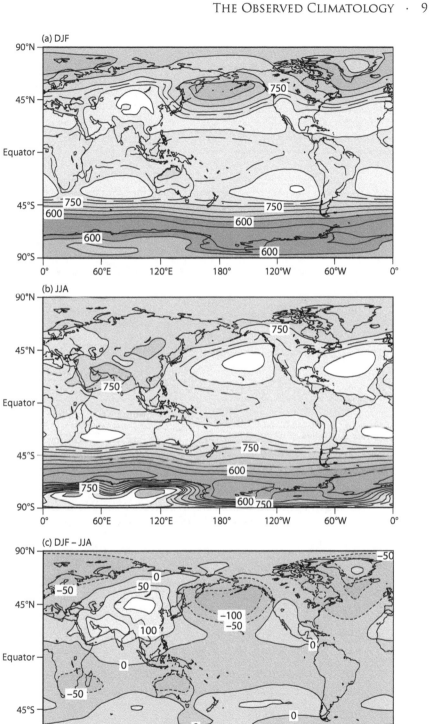

Figure 2.4 Geopotential height climatology at 900 hPa in meters for (a) December-January-February (DJF),(b) June-July-August (JJA), and (c) their difference.

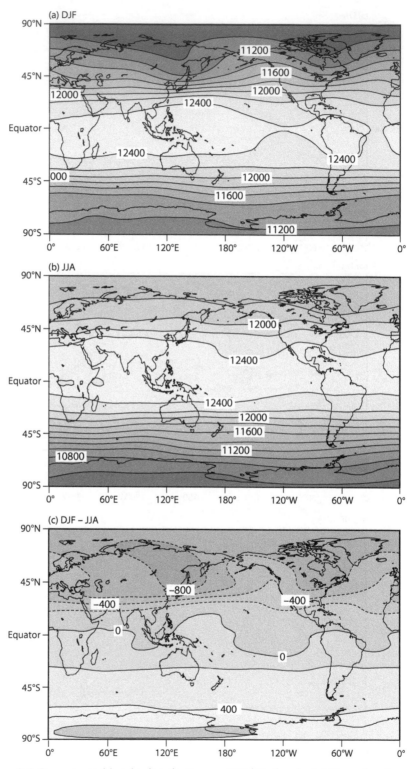

Figure 2.5 Geopotential height distributions at 200 hPa for the (a) December-January-February mean (DJF), (b) June-July-August mean (JJA), and (c) their difference.

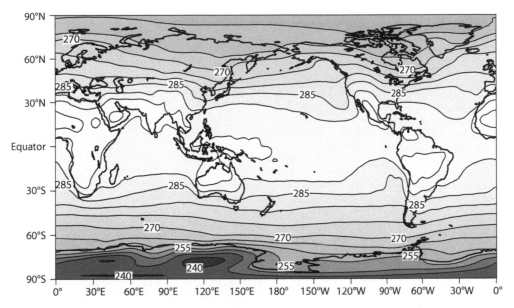

Figure 2.6 Annually averaged air temperature at 900 hPa. The contour interval is 5 K.

Figure 2.6 portrays the annual mean air temperature climatology at 900 hPa. Note the following features:

- Annual mean air temperatures over land tend to be warmer than over the oceans at the same latitude. (Note how the isotherms curve poleward over the continents in both hemispheres.)
- The lowest annual mean surface air temperatures on the planet are located over Antarctica, where they are about 60 K colder than the warmest temperatures over northern Africa.
- Surface air temperatures on the eastern sides of the Atlantic and Pacific Ocean basins are cooler than over the western sides of the basins in both hemispheres. (Note how the isotherms dip equatorward over the eastern sides of the Pacific and Atlantic Ocean basins.)
- *Continentality*, defined as the effects of land/sea distributions on climate variables, is in evidence in the temperature field, as it was for the geopotential height distributions shown above.
- The *thermal equator* marks the latitude of maximum temperature. Note that it is not necessarily located on the geographical equator and tends to favor the Northern Hemisphere.
- Meridional temperature gradients, indicated by the density of the isotherms in the north/south direction, are greater in middle and high latitudes than in the tropics.

Seasonal air temperatures at 900 hPa and their differences are displayed in Figure 2.7.

- *Seasonality*, defined here as differences in climate variables between DJF and JJA, is larger over the continents than over the oceans, and greater at high latitudes than at low latitudes.

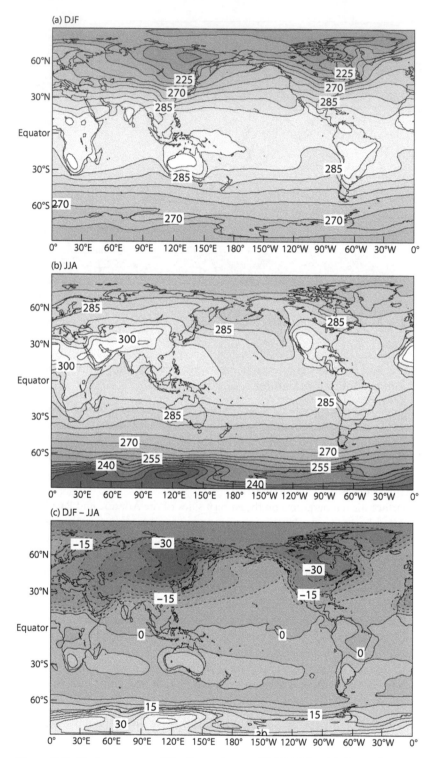

Figure 2.7 Air temperature (K) climatology at 900 hPA for (a) December-January-February (DJF), (b) June-July-August (JJA), and (c) their difference. Contour intervals are 5 K.

- Seasonality is much more pronounced in the Northern Hemisphere than in the Southern Hemisphere in association with the distribution of the continents, except over Antarctica.
- Air temperatures over land are warmer than temperatures over the oceans in the summer hemisphere, and generally cooler in the winter hemisphere. Recall from Figure 2.6 that the annual mean air temperature is warmer over land than over oceans.
- The east–west temperature gradients over the tropical oceans are in place throughout the year.

Temperature profiles provide more detail about the vertical structure of the atmosphere. Figure 2.8 shows the globally and annually averaged temperature profile through the height of the earth's atmosphere. The region of the atmosphere from the surface to roughly 12 km is termed the *troposphere*, which is derived from the Greek root meaning "the turning or changing sphere." This region is where weather systems exist and is the focus of our study of climate dynamics. Temperature decreases with height, z, in the troposphere, from a globally averaged surface air temperature of 287.5 K to about 218 K at the top of the troposphere, which is known as the *tropopause*. Pressure decreases from about 1000 hPa at the surface (the globally averaged value is generally taken as 1013 hPa) to about 200 hPa at the tropopause. Almost all the atmosphere's water resides in the troposphere, as does about 80% of its mass.

The *stratosphere* lies above the tropopause, extending to about 48 km or 1 hPa, and is capped by the *stratopause*. It is a vertically stable, stratified region—hence its name—in which temperature increases with altitude. The *mesosphere* ("middle sphere") stretches from the stratopause to about 80 km, with temperature again decreasing with height. The region of transition to interplanetary space, above 80 km, is the *thermosphere*.

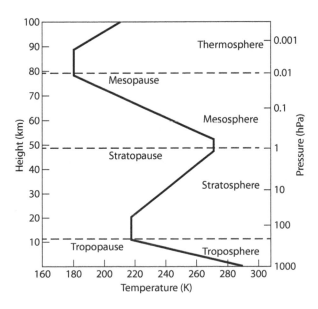

Figure 2.8 Globally and annually averaged atmospheric temperature as a function of height and pressure.

What does the temperature profile displayed in Figure 2.8 imply about the heat sources for the atmosphere? Because we expect temperature to be highest near these heat sources, Figure 2.8 suggests that the earth's surface and the stratopause are regions of heat input to the atmosphere. The troposphere is heated primarily from below. As discussed in chapter 4, much of the incoming solar radiation passes through the atmosphere and warms the surface. Heat from the surface is then delivered to the atmosphere.

The only region where the incoming solar radiation is strongly absorbed is in the stratosphere, where ozone absorbs ultraviolet wavelengths. The stratopause marks the level of maximum absorption of solar radiation by ozone, but it is not the location of the greatest ozone concentrations. Ozone concentrations generally peak at an altitude of about 25 km, but much of the ultraviolet radiation has been removed from the solar beam at that level by the ozone above.

The regions of the atmosphere are distinguished by the sign of the *lapse rate*, Γ, defined as the negative of the vertical temperature derivative[1]

$$\Gamma = -\frac{\partial T}{\partial z}.\tag{2.4}$$

The lapse rate is positive in the troposphere and mesosphere, where temperature decreases with increasing height (decreasing pressure), and negative in the stratosphere and thermosphere.

Figure 2.8 can be used to estimate the magnitude of the lapse rate between two levels (z_1 and z_2) using a finite-differenced form of Eq. (2.4):

$$\Gamma = -\frac{(T_1 - T_2)}{(z_1 - z_2)}.\tag{2.5}$$

Moving from the one-dimensional depiction of atmospheric temperature profiles in Figure 2.8, Figure 2.9 depicts atmospheric temperature as a function of latitude and pressure. Averaging in longitude completely around the globe generates the *zonal mean* plots of atmospheric temperature shown in Figure 2.9. The vertical scale is stretched relative to the horizontal scale in the figure to make the temperature structures discernible. Because the radius of the earth is 6371 km, the distance across the surface from the South Pole to the North Pole is about 20,000 km (half the circumference). The height of the troposphere and lower stratosphere is about 20 km (see Fig. 2.8). Therefore, if Figure 2.9 were drawn to scale, keeping the width of the figure the same, the height of the figure would be about 0.005 in. In other words, the horizontal scale of the atmosphere is much greater than its vertical scale. In this sense, the atmosphere is thin.

Figure 2.9 provides an opportunity for quantifying meridional variations in temperature. In spherical coordinates, and using finite differencing, the zonal mean meridional temperature gradient is

$$\frac{1}{a}\frac{\partial [T]}{\partial \phi} \to \frac{1}{a}\frac{(T_1 - T_2)}{(\phi_1 - \phi_2)},\tag{2.6}$$

---

[1] Note that the partial derivative, $\partial/\partial z$, is used instead of the total derivative, d/dz, because temperature depends on other independent variables (e.g., latitude and longitude) as well as on z.

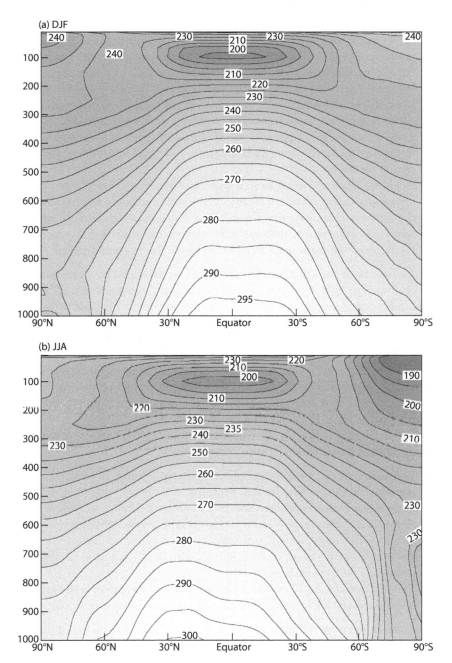

Figure 2.9 Zonal mean temperature climatology for (a) December-January-February (DJF) and (b) June-July-August (JJA). Contour interval is 5 K.

where the square brackets indicate the zonal mean, a is the radius of the earth, and ϕ is latitude in radians.

Note the following in Figure 2.9:

- The tropopause, where the lapse rate changes sign, is located near 100 hPa deep in the tropics and comes closer to the surface in higher latitudes.

- There is no tropopause at very high latitudes in the winter hemisphere. In the absence of shortwave (ultraviolet) radiation to absorb, ozone does not provide a heat source and temperatures continue to decrease with elevation.
- Polar temperature inversions, in which temperatures increase with height (decreasing pressure), occur near the surface at high latitudes in winter. Again, in the absence of insolation, the main heat source for the polar atmosphere in winter is the atmospheric transport of warm air from lower latitudes.
- Zonal mean meridional temperature gradients are larger in the winter hemisphere than in the summer hemisphere, and larger in middle latitudes than in the tropics.
- Near the surface, the zonal mean temperature maximum is very close to the equator during Northern Hemisphere winter and moves farther into the Northern Hemisphere (to about 25°N) during Northern Hemisphere summer. In other words, the zonal mean temperature distribution is more symmetric about the equator in DJF than in JJA due to continentality.

Information about the atmospheric circulation is provided by examining wind distributions. The east–west component of the wind is referred to as the *zonal wind*, and it is denoted by u. The sign convention is that u is positive when the wind is blowing from west to east, that is, in the same direction as the rotation of the earth, and negative when the flow is from east to west. The term *easterly* means "from the east," so it is a synonym for westward. Similarly, *westerly* means "from the west," or eastward. The north–south component of the wind is called the *meridional wind*, denoted by v. Positive values of v denote a *southerly* wind, that is, flow from the south to the north, while negative values of v denote a *northerly*, or southward, wind.

Figure 2.10 depicts the zonal mean zonal wind, or the east–west wind averaged around all longitudes.

- In the troposphere, the flow is primarily westerly (positive) in middle latitudes and easterly (negative) in the tropics.
- In the midlatitude troposphere, winds become more strongly westerly with elevation until they reach a maximum at the tropopause. These maxima are known as the *subtropical westerly jets*.
- The tropical easterly flow is more uniform with elevation and so has less *wind shear* than the midlatitude flow.
- The subtropical jets are stronger in the winter hemisphere than in the summer hemisphere, and are located 15° of latitude closer to the equator in winter.

The zonal mean meridional, or north–south, wind distribution is shown in Figure 2.11. Here, the contour interval used is 0.5 m/s, an order of magnitude smaller than the contour interval that was used to display the zonal mean zonal wind in Figure 2.10. Arrows indicate the wind direction.

According to Figure 2.11:

- The greatest values of the zonal mean meridional wind are found in the upper troposphere (300–100 hPa) in the tropics, and near the surface (below 800 hPa).
- The low-level flow in the tropics converges in the meridional direction. The centerline of this *convergence* is located in the summer hemisphere. During

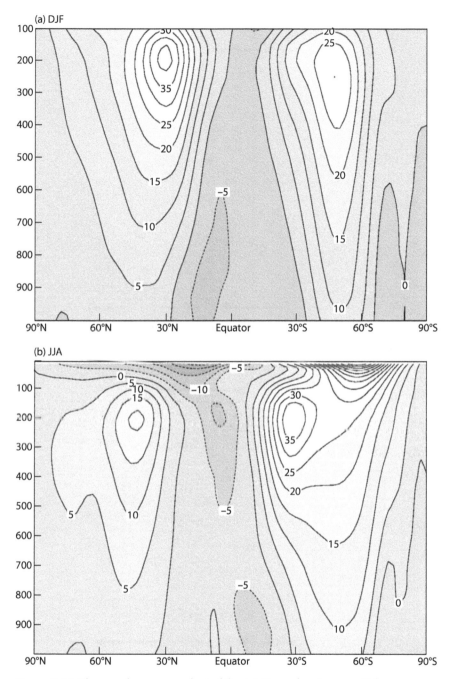

Figure 2.10 The zonal mean zonal wind for (a) December-January-February (DJF) and (b) June-July-August (JJA). Contour interval is 5 m/s and negative values denote easterly flow.

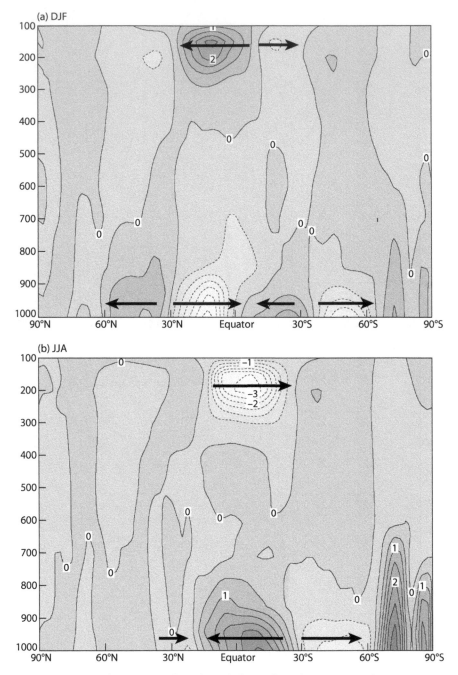

Figure 2.11 Zonal mean meridional wind climatology for (a) December-January-February (DJF) and (b) June-July-August (JJA). Contour interval is 0.5 m/s and negative values denote northerly flow. Arrows indicate wind direction.

Southern Hemisphere summer (DJF), the center of the convergence is only a few degrees from the equator, but during Northern Hemisphere summer (JJA) it is located near 20°N.

- The upper-level flow in the tropics is divergent in the meridional direction. The centerline of this *divergence* is also located in the summer hemisphere and, as for the low-level convergence, the center is close to the equator in DJF and well into the Northern Hemisphere in JJA.

When pressure is used as a vertical coordinate instead of elevation, the vertical velocity, ω, is expressed in units of pascals per second (Pa/s). This variable is known as the vertical *p-velocity* and defined as

$$\omega \equiv \frac{dp}{dt}. \tag{2.7}$$

A negative value of ω indicates that a volume of air is moving such that pressure is decreasing with time. Because pressure decreases with elevation (Fig. 2.2), this means that negative values of ω denote upward velocity.

Figure 2.12 shows the zonal mean vertical *p*-velocity climatology for DJF and JJA. The C's in the figure denote the approximate location of the meridional wind convergence at low levels based on Figure 2.11. Upward motion through the depth of the troposphere occurs over these regions of convergence. Similarly, deep sinking motion is associated with low-level divergence (denoted by the D's in the figure). It is important to recall that these upward and downward motions are averages over long time periods and over large space scales. Locally, the flow may be quite different.

The patterns of wind direction and magnitude observed in the zonal mean zonal wind (Fig. 2.10) and the zonal mean meridional wind (Fig. 2.11) at 900 hPa are combined in a map of wind vectors, shown in Figure 2.13. Geopotential height lines are also drawn for reference since, as will become clear in following chapters, there is a close relationship between the flow and the geopotential height distribution.

Note the following:

- In the tropics, low-level winds are generally northeasterly in the Northern Hemisphere and southeasterly in the Southern Hemisphere.
- The meridional flow converges near the equator, consistent with Figure 2.11. This region is called the *Intertropical Convergence Zone (ITCZ)*.
- The monsoons (e.g., over South America in Southern Hemisphere summer and southeastern Asia in Northern Hemisphere summer) cause much of the east/west asymmetry in the low-level flow field. Note the 180° change in the wind direction with season off the west coast of India in the Arabian Sea and over the Bay of Bengal off the east coast.
- Outside the deep tropics, the wind blows parallel to the geopotential height lines.
- The flow around regions of low geopotential heights is counterclockwise in the Northern Hemisphere and clockwise in the Southern Hemisphere. This is termed *cyclonic* flow.

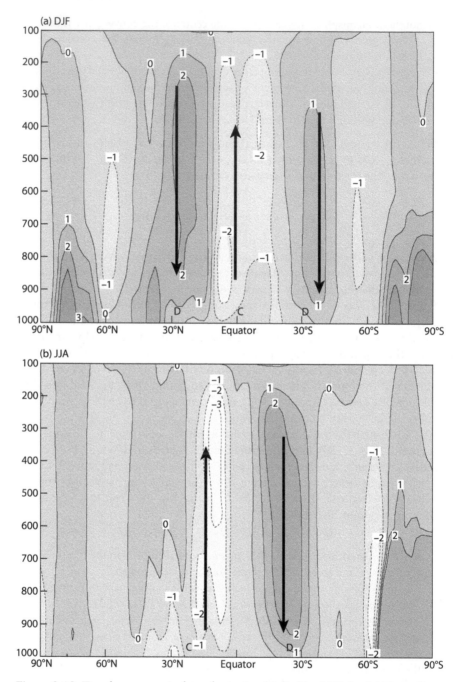

Figure 2.12 Zonal mean vertical *p*-velocity (multiplied by 100) for (a) December-January-February (DJF) and (b) June-July-August (JJA). Contour interval is 0.01 Pa/s and negative values denote upward motion. Arrows indicate wind direction.

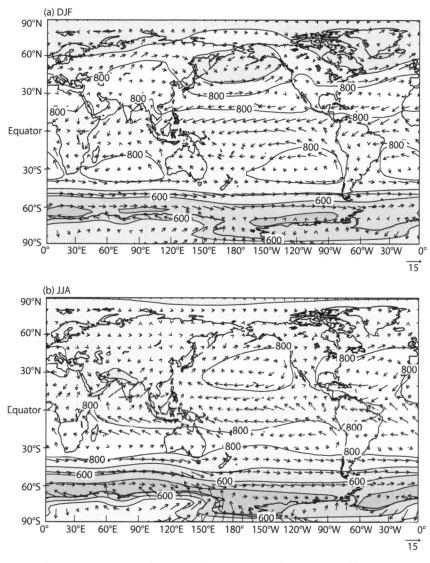

Figure 2.13 Lower-troposphere (900 hPa) winds and geopotential height contours for (a) December-January-February (DJF) and (b) June-July-August (JJA). The vector scales indicated in the lower right are in m/s.

- The flow around regions of high geopotential heights is clockwise in the Northern Hemisphere and counterclockwise in the Southern Hemisphere, and this is known as *anticyclonic* flow. The reasons for these flow patterns will become clear when we study the forces that produce them in chapter 6.

Winds in the upper troposphere, represented by the 200 hPa level in Figure 2.14, are considerably more zonally uniform than those in the lower troposphere, consistent with the geopotential height structure. Especially easy to see in this figure is that zonal wind speeds are greater where geopotential height

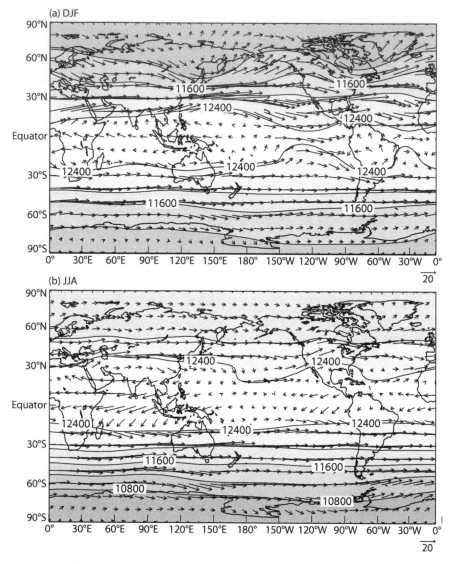

Figure 2.14 Upper-troposphere (200 hPa) winds and geopotential height contours for (a) December-January-February (DJF) and (b) June-July-August (JJA). The vector scale indicated in the lower right is in m/s.

lines pinch together, that is, where meridional geopotential height gradients are larger. This happens most prominently in DJF in the storm tracks off the east coasts of Asia and North America.

2.2   THE OCEAN

Observations of the oceans are less complete than observations of the atmosphere. Oceanographers traditionally relied, in part, on "ships of opportunity" for direct measurements of surface water temperatures and currents. These were merchant ships whose crews voluntarily made observations of the upper oceans and reported them to a central location. Besides using nonstandardized recording methods, these ships provided information only from shipping lanes, leaving large portions of the ocean surface unobserved.

Oceanographers have been creative in developing various ways of retrieving information about the oceans. In May 1990, for example, they took advantage of a spill of 60,000 sports shoes from a cargo ship in the North Atlantic. In another incident 29,000 bathtub toys were spilled in a storm near the date line in 1992. When the spilled items started turning up along the North American coast several months later, oceanographers used their locations to infer speeds and directions of ocean currents.

The satellite era for earth observations, which began with the launch of *TIROS-1* in April 1961, brought global observing coverage of the ocean's surface. Sea surface temperature, surface winds, phytoplankton distributions, sea surface elevations, and other variables describing the state of the ocean's surface are extracted from satellite observations.

Observing the ocean away from the surface presents a formidable challenge because the world's oceans are enormous and, for the most part, remote. Ocean circulation patterns and thermal structure beneath the surface are primarily inferred from indirect measurements or from drifters. Direct measurements of ocean temperature, pressure, salinity, and density are either the result of extrapolating in space from local measurements scattered around the globe, or extrapolating in time from intensive observing periods. Thus, we have less confidence in our climatological picture of the ocean than the atmosphere, and we have only a rudimentary idea of how the ocean's climate varies in time.

The heat capacity of the ocean dwarfs that of the atmosphere. There is more heat energy in the top 3 m of the ocean than in the entire atmospheric column above it. The specific heat of water, defined as the amount of heat needed to increase the temperature of 1 kg of water by 1 K, is about 4218 J/(kg · K), and varies slightly with temperature and pressure. This is four times the specific heat of air, which is about 1007 J/(kg · K), again depending on temperature and pressure. Furthermore, the mass of the atmosphere is minuscule compared with that of the oceans (5.3×10^{18} kg versus 1.4×10^{21} kg, respectively). Consequently, the heat content of the world's oceans is immense compared with that of the atmosphere, and sea surface temperature distributions are very powerful forcing functions for the atmosphere.

Figure 2.15 shows the sea surface temperature climatology, annually averaged for the period 1900–1997. (Note that the zonal boundaries of the plot are shifted eastward by 20° of longitude compared with the figures of the atmosphere in the previous section so the Atlantic Ocean basin is not disrupted.) Regions over the oceans with no data are covered by sea ice for at least some portion of the year. Note the following features:

- Meridional temperature gradients are larger in middle and high latitudes than in the tropics.
- Sea surface temperatures in the eastern sides of the Atlantic and Pacific Ocean basins are cooler than in the western sides of the basins in both hemispheres, similar to the structure seen in the low-level atmospheric temperature (Fig. 2.6). The western tropical Pacific, known as the *western*

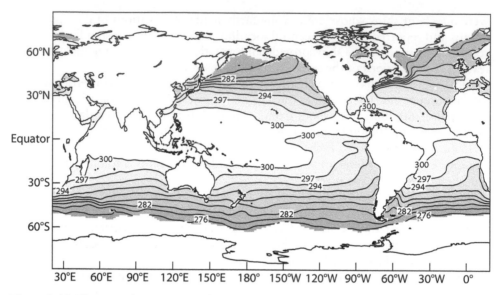

Figure 2.15 The annual mean sea surface temperature climatology. Contour interval is 3K.

warm pool, is more than 5 K warmer than the eastern tropical Pacific, which is referred to as the *cold tongue* region.

- In contrast to the Pacific and Atlantic Oceans, annual mean Indian Ocean sea surface temperatures are zonally uniform.
- Sea surface temperature gradients are especially strong off the east coasts of Asia and North America, north of about 30°N, similar in placement to the atmospheric geopotential height gradients and temperatures in this region.

Sea surface temperatures for DJF and JJA are displayed in Figure 2.16, along with their difference.

- The greatest seasonality in sea surface temperatures occurs in the middle latitudes of the North Pacific and North Atlantic Oceans.
- Seasonal differences in sea surface temperatures are much smaller than seasonal differences in surface air temperatures (compare with Fig. 2.7c).
- Southern Hemisphere seasonality in sea surface temperatures is smaller and more disorganized than seasonality in the Northern Hemisphere.

To explore the vertical dependence of ocean temperature, meridional cross sections taken through the Pacific and Atlantic Oceans are displayed in Figure 2.17. The Pacific Ocean section is along the International Date Line (180° longitude), and the Atlantic Ocean section is along 30°W. The white regions outline the ocean bottom topography, and where they reach the surface they indicate the presence of continents. Note the distortion in the space scale. As for the atmospheric cross sections, the vertical scale is expanded in the figure relative to the horizontal scale so the details of the temperature structure can be seen. But again, as for the atmosphere, the vertical scale of the ocean is many orders of magnitude smaller than its horizontal scale.

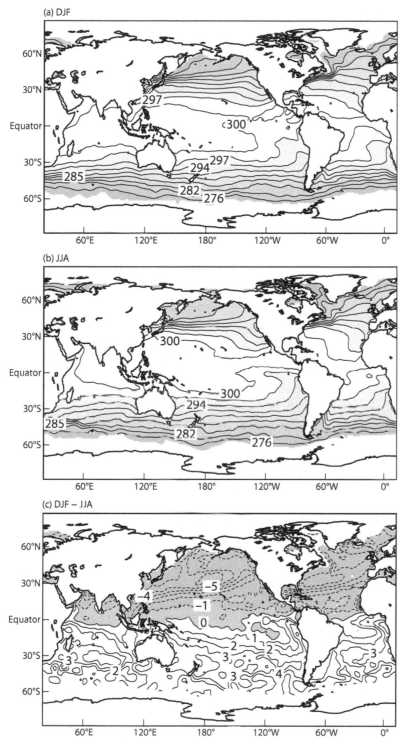

Figure 2.16 Sea surface temperatures (K) for (a) December-January-February (DJF), (b) June-July-August (JJA), and (c) their difference. Contour interval is 3K in (a) and (b), and it is 1K in (c). Negative values are shaded in (c).

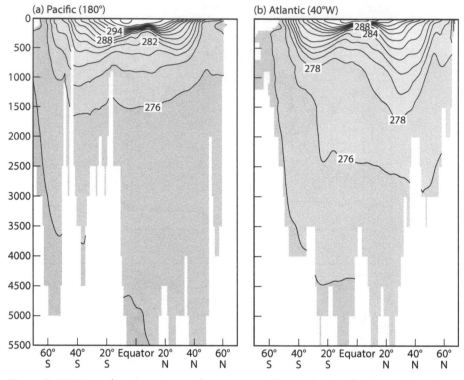

Figure 2.17 Vertical cross sections of temperature for (a) the Pacific Ocean along the International Date Line and (b) the Atlantic Ocean at 30°W. Contour interval is 2 K, and depth is in m.

Note the following in Figure 2.17:

- The warmest waters, as you would expect, are located at the surface in the tropics.
- Between 40°S and 40°N, the isotherms are aligned horizontally, denoting the stable layering of the oceans at these latitudes, with warmer (less dense) water floating on top of cold water.
- Isotherms at high latitudes in the Southern Hemisphere, and high latitudes in the North Atlantic (but not in the North Pacific), are oriented vertically. In these regions, surface waters are as cold as the deep ocean water, just a few degrees above the freezing point of fresh water (273.15 K). This orientation of the isotherms at high latitudes provides clues about the large-scale circulation of the oceans, discussed in chapter 9.

Three layers are defined for the ocean. The *ocean mixed layer* (or simply "the mixed layer") is characterized by strong interactions with the atmosphere. Surface winds induce robust vertical mixing and horizontal currents in this region. As a consequence of this stirring of the ocean by the atmosphere, temperature is not a strong function of depth in the mixed layer. Depending on season and latitude, the mixed layer is 20–200 m deep and contains roughly 2% of the volume of the ocean. The permanent *thermocline* (or *pycnocline*) is located between 200 and 800 m depth, depending on the geographic location. It is stably stratified, with temperature decreasing and density increasing with depth. The deep ocean, which extends to the bottom below the permanent thermocline, contains 80% of the ocean's water. Here, temperature decreases slowly with depth.

Additional insight about the thermal structure of the upper layers in the tropics results from taking east–west cross sections through the Atlantic (50°W to 10°E) and Pacific (130°E to 80°W) Oceans along the equator. The mixed layer and the upper thermocline are represented in Figure 2.18.

- The mixed layer in the west in both basins is relatively deep, or well developed, with warm water extending to 100 m depth or more. (Note the weak stratification in this region denoted by the wide spacing of the isotherms.)
- The depth of the mixed layer decreases to the east, and the thermocline tilts upward and brings cooler water to the surface in the eastern ocean basins. This structure is consistent with Figure 2.15, which indicates that sea surface temperatures in the western parts of these basins are significantly warmer than in the east.
- Below about 250 m depth, the isotherms are flat.

Salinity is related to the mass of dissolved salts in 1 kg of seawater. Most of the dissolved salts contain sodium and chloride ions, but magnesium, sulfur, calcium, and potassium ions are also present. Salinity values can be written in terms of grams of salt per kilogram of water (g/kg), or parts per thousand (ppt).

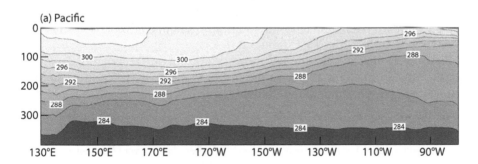

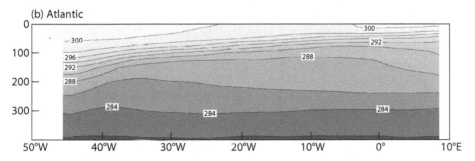

Figure 2.18 Cross sections of annual mean ocean temperature (K) in the upper 400 m of the ocean along the equator for (a) the Pacific (130°E to 80°W) and (b) the Atlantic (50°W to 10°E) Ocean basins. White areas are continents.

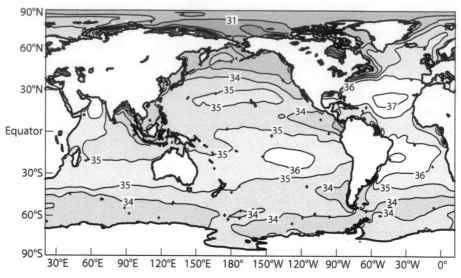

Figure 2.19 Annual mean sea surface salinity. The contour interval is 1 psu.

Oceanographers also define salinity in terms of the conductivity of a sample of water, since a higher conductivity is associated with the presence of more ions and dissolved salts. The *practical salinity unit* (psu) is the standard measure of salinity, although it is technically not a unit because it is dimensionless. Values indicate the ratio of the conductivity of a sample of seawater to the conductivity of a standard potassium chloride solution. A salinity value of 35 psu is very close to 35 g/kg of salt at 15°C (about 258 K).

Figure 2.19 displays the annual mean salinity of surface waters. *Isohaline* values range from about 29 psu in the Arctic to more than 37 psu in the subtropical Atlantic Ocean. Note the following:

- The Atlantic Ocean is saltier than the Pacific and Indian Oceans.
- Salinity tends to be highest in the subtropics.
- Lower salinity values occur near the coasts where large rivers flow into the sea. Examples are the northern Gulf of Mexico near the mouth of the Mississippi River, and off the northeast coast of South America where the Amazon and Orinoco Rivers empty into the Atlantic.
- High salinity gradients are located off the east coast of North America, north of about 30°N, similar to the strong gradients in temperature in this region (Fig. 2.15).
- Atlantic waters with high salinity extend far north into the Arctic east of Greenland.

The latitudinal dependence of surface water salinity is apparent in the zonal mean plot in Figure 2.20. In the global average, subtropical surface salinity is about 1.5 psu greater than that close to the equator. The black line in the figure is the zonal mean excess of evaporation (E) over precipitation (P), illustrating the association of their difference (E − P) with ocean surface salinity. When water

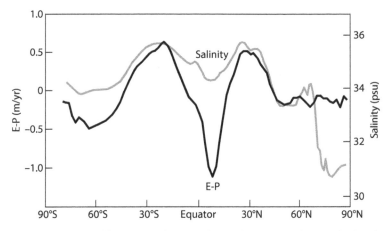

Figure 2.20 Zonally averaged sea surface salinity (gray line) calculated for all oceans and the difference between evaporation and precipitation (E – P; black line).

evaporates from the ocean, the salt left behind increases surface salinity. Subtropical waters are relatively salty because E > P in this region. Near the equator, in the ITCZ, P > E and the surface waters are freshened. The relationship breaks down at high latitudes. In the Arctic, surface waters are freshened by outflow from numerous rivers. The high salinity of the Antarctic circumpolar region is a result of sea ice formation, which leaves salt behind in the ocean.

Figure 2.21 shows the vertical and meridional structure of salinity in the Atlantic and Pacific Ocean basins. Highest salinity values do not occur at the surface, but near the base of the ocean mixed layer. The high salinity of the subtropical Atlantic Ocean, especially in the Northern Hemisphere, is displayed clearly in the figure. The salinity of polar surface waters is relatively low, and the vertical alignment of the contours at high latitudes is similar to that of the isotherms in Figure 2.17.

Ocean currents are analogous to wind in the atmosphere for portraying circulation. Figure 2.22 is a schematic representation of ocean surface current directions.

- *Western boundary currents* flow along the western edges of each ocean basin. Examples are the Gulf Stream, the Kuroshio, the Agulhas Current, and the Brazil Current. These are warm currents, so called because they carry relatively warm water from lower latitudes to higher latitudes.
- *Eastern boundary currents* flow along the eastern edges of each ocean basin. Examples are the California, Peru, and Benguela Currents. These are cold currents that flow from high latitudes to low latitudes in both hemispheres.
- Boundary currents are stronger and more well developed in the Northern Hemisphere, because the continents in that hemisphere enclose the ocean basins more completely.
- The North and South Equatorial Currents straddle the equator and flow from east to west. The Equatorial Countercurrent lies in between, and it brings water back eastward below the surface.

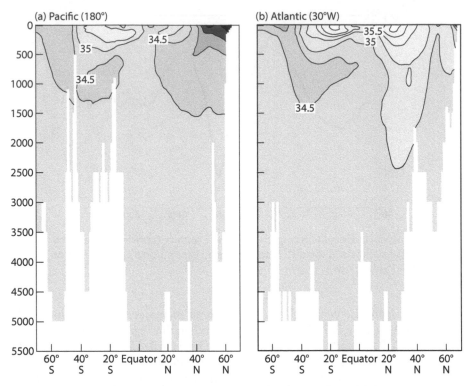

Figure 2.21 Annual mean salinity cross sections for the Pacific and Atlantic Oceans taken (a) along the International Date Line (180°) and (b) at 30°W. The contour interval is 0.5 psu. From *World Ocean Atlas*, Conkright et al. (1998).

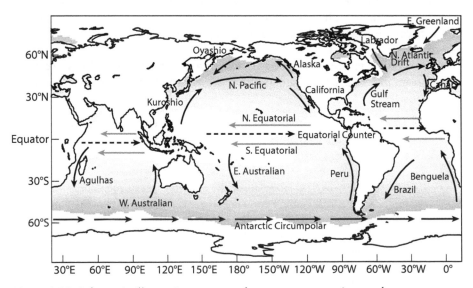

Figure 2.22 Schematic illustrating ocean surface currents superimposed over sea surface temperature contours as in Figure 2.15.

- The Antarctic Circumpolar Current, or the "West Wind Drift," is the only current that circles the globe. This is possible because there is no land at any longitude between about 50°S and 60°S. The water must all be funneled through the narrowest point, the Drake Passage, at the tip of South America.

The currents within an ocean basin can be connected into *gyres*, which are roughly circular current systems that flow anticyclonically in each ocean basin. For example, the North Pacific gyre consists of the Kuroshio, North Pacific, California, and North Equatorial Currents. The western boundary currents within the gyres are faster, narrower, and deeper than the eastern boundary currents.

Note that the directions of the surface currents are similar to lower-tropospheric wind directions (Fig. 2.13). In the tropics, off the equator, the surface currents are easterly, and in the middle latitudes the surface currents are westerly except close to the coasts. The boundary currents close the circulation within each ocean basin. Surface winds are 5–15 m/s on average (Fig. 2.13), but current speeds are only a few percent of the low-level wind speed. The Gulf Stream velocity, for example, is 15–20 cm/s in subtropical and middle latitudes, and its velocity drops to roughly 2–4 cm/s as it turns eastward to cross the Atlantic.

The influence of the boundary currents can be seen in sea surface temperature distributions (Figs. 2.15 and 2.16). For example, the Peru Current in the eastern South Pacific Ocean basin transports cooler surface waters equatorward and helps maintain the cold tongue structure and, thereby, the longitudinal temperature gradient that is characteristic of the tropical Pacific Ocean. The relatively warm waters of the eastern Atlantic Ocean in northern midlatitudes are caused by the northward transport of warm tropical water by the Gulf Stream.

There are similar correlations between surface currents and the surface salinity distribution (Fig. 2.19). The western North Atlantic basin is saltier than the eastern basin because the Gulf Stream transports salty water of Mediterranean origin northward. Other examples are the freshening of the eastern portions of the North Pacific and South Atlantic basins by the California and Benguela Currents, respectively.

In addition to the essentially horizontal surface currents (Fig. 2.22), the ocean has a coherent global-scale three-dimensional circulation. While ocean surface currents are largely driven by surface winds, the three-dimensional circulation is driven by density differences. It is known as the *thermohaline circulation* because both temperature and salinity influence seawater density (see chapter 5).

Figure 2.23 illustrates this global circulation system, known as the *great ocean conveyor belt*. Many details of the large-scale ocean circulation are not known, since the volume of the ocean to be monitored is immense and difficult to reach, for the most part. In addition, the circulation is slow and the time scales are very long—upward of 1000 years for the global-scale ocean circulation—so a few decades of observations will not reveal the full scope of the motion. (Some of the techniques used by oceanographers to piece together a picture of the ocean's global circulation system are discussed in chapter 8.)

According to the ocean conveyor belt model (Fig. 2.23a), water sinks in the North Atlantic into the deep ocean (~2–4 km). This is known as the *North*

(a)

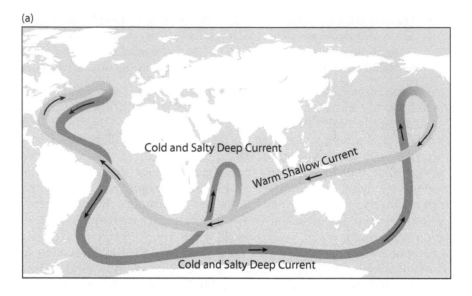

(b)

(c)

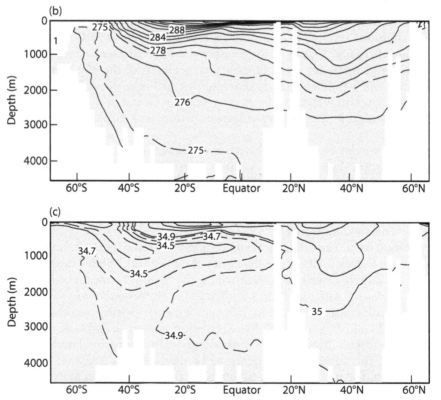

Figure 2.23. (a) Simplified schematic illustration of the thermohaline circulation. (b) Temperature and (c) salinity are shown in cross-sections in the Atlantic Ocean at 18°W. Warm, salty Mediterranean water is evident at 1500m depth near 35°N. North Atlantic water sinks to intermediate depth (about 2500m) and overlays colder but fresher Antarctic bottom water.

Atlantic deep water formation, and its presence is indicated by the high-latitude vertical isotherms and isohalines in Figures 2.23b and c. The North Atlantic deep water flows south to the Antarctic circumpolar region, where it joins the westerly flow of the Antarctic circumpolar circulation. Water sinks to even greater depths (~4–5 km) in this region, forming the *Antarctic bottom water*. The deep and bottom waters of the Antarctic circumpolar westerly flow upwell to the surface in either the Indian Ocean or the subtropical and tropical Pacific. The upper-ocean return flow that balances the sinking in the North Atlantic consists of the low-latitude easterly flow from the Pacific, past Indonesia (known as the *Indonesian Throughflow*) into the Indian Ocean, around the southern tip of Africa into the South Atlantic Ocean. Approximately 20% of the mass flux of the Gulf Stream is thought to be associated with the thermohaline circulation, with the remainder wind driven.

Another facet of the ocean circulation that is important for climate is the upwelling of cold water and downwelling of warm water. Upwelling can be wind driven or mixing driven, the latter in association with the thermohaline circulation. Wind-driven upwelling occurs when the wind blows warm surface water away from a coast and cooler waters rise as a result. Downwelling occurs when the wind blows toward a coast. Upwelling also takes place when the wind blows along a coast and, driven by Coriolis accelerations (discussed in chapter 6), surface waters flow away from the coast (see chapter 8). One example of this type of wind-driven upwelling occurs off the east coast of Africa near Somalia, where the low-level southerly wind (Fig. 2.13b) forces the ocean surface waters to flow eastward because of Coriolis acceleration. Upwelling of colder water results, which accounts for the cooler sea surface temperatures in the western Indian Ocean during Northern Hemisphere summer (Fig. 2.16). Similar upwelling occurs off the Peru and California coasts, and in the Gulf of Guinea (south of Africa's westward bulge). The nutrients brought closer to the surface by coastal upwelling attract marine life and support the fishing industries of many countries.

2.3 THE HYDROLOGIC CYCLE

Water cycles among the interdependent components of the climate system and is an important agent in producing interactions among them (Fig. 1.1).

The volume of water in the *hydrosphere*, defined as all the water in all phases (solid, liquid, and vapor) in the earth system down 2 km into the crust, is estimated to be 1.4×10^{10} km^3. Approximately 97.5% of this water is saline, so only 2.5% is fresh water. Table 2.1 gives the estimated distribution of water in various reservoirs.

Estimated annual mean global fluxes and reservoirs of water in the climate system are shown in Figure 2.24. Over land, precipitation is greater than evaporation or, more properly, *evapotranspiration*, by about 36×10^{15} kg/year. The opposite is true over the oceans, where evaporation rates exceed precipitation rates. Note that although the oceans cover approximately 70% of the earth's surface, 86% of the total evaporation is from the ocean. The atmosphere transports the excess water evaporated from the oceans to the land, indicating the

Table 2.1 Distribution of water in the climate system

Location	Percentage of global water	Volume of water (km^3)	Mass of water (kg)
Oceans	97%	1.37×10^{10}	1.37×10^{22}
Ice (glaciers, sea ice)	2%	2.9×10^{8}	2.9×10^{20}
Groundwater	0.7%	9.5×10^{7}	9.5×10^{19}
Lakes	1×10^{-2}	1.25×10^{6}	1.25×10^{18}
Soils	5×10^{-3}	6.5×10^{5}	6.5×10^{17}
Atmosphere	1×10^{-3}	1.3×10^{5}	1.3×10^{17}
Rivers and streams	1×10^{-4}	1.7×10^{4}	1.7×10^{16}
Biosphere	4×10^{-5}	6×10^{3}	6×10^{15}

important role of the atmospheric circulation in the global hydrologic cycle. The climatological water cycle is completed when surface and underground runoff transport water from the land to the oceans.

Water cycles rapidly through some components of the climate system, and much more slowly through others. The average residence time for a water vapor molecule in the atmosphere is only about 1 week. Residence times for water in other parts of the climate system range from days in soils and rivers, to many decades in glaciers and lakes, to 10,000 years or more in deep groundwater.

Precipitation is highly variability in space and time, making it difficult to observe and requiring a long averaging time to form a reliable climatology. While excellent rainfall records have been kept in some regions—such as India,

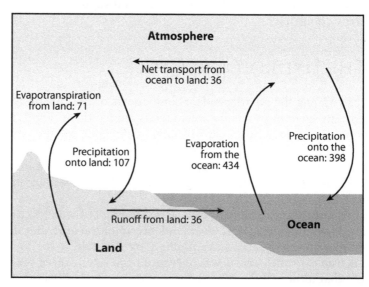

Figure 2.24 Estimated climatological fluxes (in 10^{15} kg/yr) of water in the climate system.

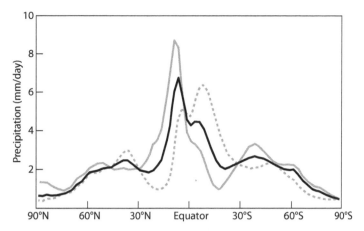

Figure 2.25 Zonally averaged precipitation climatology in mm/day for the annual mean (black line), December-January-February (DJF) mean (dotted line), and June-July-August (JJA) mean (gray line).

China, and Europe—for centuries, other regions—such as the South Pacific Ocean—have very few direct measurements of precipitation available. The satellite era has produced global precipitation data sets, but there are still uncertainties associated with translating the radiative fluxes measured by satellites into precipitation and with producing adequate global coverage.

Figure 2.25 displays zonal mean precipitation rates for DJF, JJA, and the annual average. The values are from a global precipitation climatology (1979–2011) developed at the National Oceanic and Atmospheric Administration's (NOAA's) Climate Prediction Centers (CPC)[2] by combining measurements from five satellites with rain gauge measurements made at weather stations and on ships around the world. Note the following:

- High precipitation rates near the equator mark the ITCZ, where the low-level winds converge (see Figs. 2.11 and 2.13). Tropical precipitation rates are about three times those in middle latitudes.
- Precipitation rates are not at a maximum on the equator. Rather, the highest precipitation rates in each hemisphere tend to occur about 7° of latitude away from the equator.
- The summer tropical precipitation maximum in the Northern Hemisphere is significantly larger than that in the Southern Hemisphere.
- Precipitation rates have a local minimum in the subtropics.
- Secondary precipitation maxima occur in middle latitudes, between 35° and 55°, depending on season.
- Lowest precipitation rates are found in high latitudes. With precipitation rates under 0.5 mm/day, it takes many hundreds of years to accumulate snow and ice in these regions.

[2] This data set is known as the CPC Merged Analysis of Precipitation (CMAP).

Figure 2.26 displays annual mean precipitation across the globe from the CMAP climatology. Note the following:

- There is pronounced east–west structure in the precipitation fields. Latitude is not necessarily a good indicator of rainfall rates, even within the ITCZ.
- The eastern sides of the Pacific and Atlantic Ocean basins are significantly drier than the western sides in both hemispheres. The opposite is true in the Indian Ocean north of about 15°N.
- In middle and high latitudes, land surfaces generally receive less rainfall than ocean surfaces.
- In both hemispheres, regions of high rainfall extend diagonally off the east coasts of the continents, to the southeast in the Southern Hemisphere and to the northeast in the Northern Hemisphere. These are the *land-based convergence zones*. In the Southern Hemisphere, these regions are known as the South Pacific Convergence Zone (SPCZ), the South Atlantic Convergence Zone (SACZ), and the South Indian Convergence Zone (SICZ).

Seasonality in global rainfall distributions is displayed in Figure 2.27.

- In midlatitudes, rainfall is concentrated over the oceans in winter and tends to be more continental in summer.
- The subtropical precipitation minimum seen in the zonal mean (Fig. 2.25) is associated with low rainfall rates over the eastern Pacific and Atlantic Ocean basins, and low precipitation rates over northern Africa (the Sahara Desert) and the Middle East, as well as parts of Asia (the Gobi Desert) and Australia.
- Tropical precipitation rates are greater in the Northern Hemisphere than in the Southern Hemisphere for two main reasons. One is that the intense precipitation band across the tropical Pacific just north of the equator (ITCZ) remains in place all year around. The other is the high precipitation rates, and great extent, of the summer monsoon in southeast Asia.
- The SACZ and the SICZ are features of the summer season. The SPCZ is present during both summer and winter, but it is better defined during the summer.

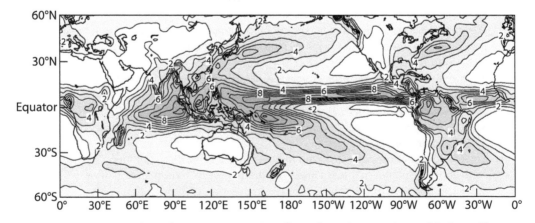

Figure 2.26 Annual mean precipitation climatology. Contour interval is 2 mm/day.

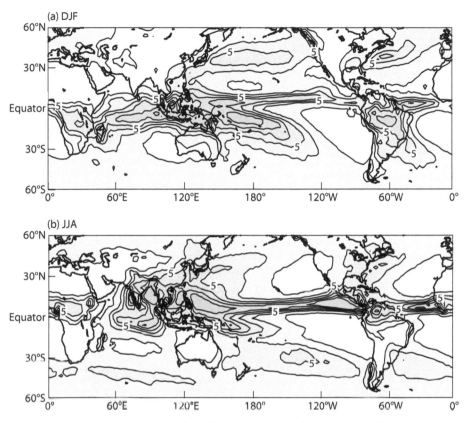

Figure 2.27 Precipitation climatology for (a) the December-January-February (DJF) mean and (b) the June-July-August (JJA)mean. Contour interval is 2 mm/day.

Evaporation rates are very difficult to measure, and we currently do not have a global climatology compiled from direct observations. Figure 2.28 shows estimated zonal mean evaporation rates from a reanalysis. Compare this figure with the zonal mean precipitation rates shown in Figure 2.25. Note that the greatest values of evaporation rates do not occur at the same latitudes as those of precipitation. Instead, they are largest in the subtropics where precipitation rates are relatively low and they decrease fairly uniformly toward the poles. The local evaporation minimum near the equator is co-located with the surface wind speed minimum (Fig. 2.10).

The geographic distribution of annual mean evaporation is displayed in Figure 2.29. Note that, in general, the evaporation distribution reflects the structure of the surface temperature field (Fig. 2.6) much more closely than it reflects the structure of the precipitation field (Fig. 2.26). An exception is very close to the equator, where the low surface wind speeds of the ITCZ (the "doldrums"; see Figs. 2.10 and 2.11) inhibit evaporation despite warm temperatures.

Specific humidity, q, is a variable commonly used to quantify the amount of water vapor in the atmosphere. Specific humidity is defined as the ratio of the mass of water vapor to the total mass (dry air plus water vapor) of a volume of

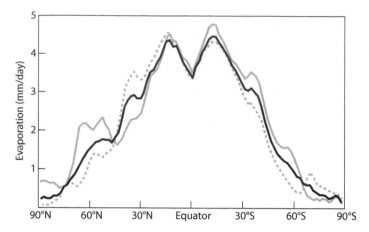

Figure 2.28 Zonally averaged evaporation climatology in mm/day
for the annual mean (black line), December-January-February mean
(dotted line), and June-July-August mean (gray line).

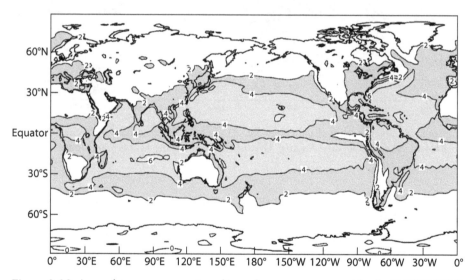

Figure 2.29 Annual mean evaporation climatology as represented in the NCEP/NCAR
reanalysis. Contour intervals are 2 mm/day.

air. Direct measurements of specific humidity are not available for constructing a global climatology, so reanalysis is used to provide an estimate of how
water vapor is distributed in the atmosphere. The zonal mean specific humidity
distribution (Figure 2.30) reveals the following:

- Most of the water vapor in the atmosphere is located close to the surface,
  within the lowest 1 or 2 km of the troposphere.
- Specific humidity decreases approximately linearly with pressure up through
  the troposphere (i.e., to first order, the spacing between the contour lines is
  constant). Because pressure falls off exponentially with height, this implies

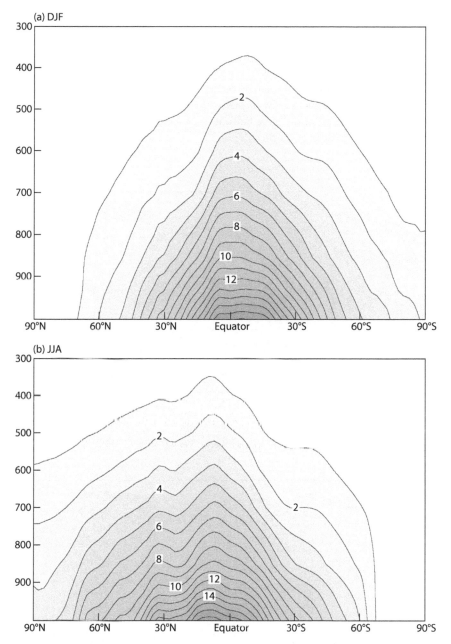

Figure 2.30 Zonal mean specific humidity climatology for (a) December-January-February (DJF) and (b) June-July-August (JJA). Units are 10^{-3} kg_{H_2O}/kg_{air}.

that specific humidity also decreases exponentially with height, to first-order approximation.

- Values are largest deep in the tropics, and decrease toward both poles.
- The summer hemisphere is more moist than the winter hemisphere, and seasonal differences are more pronounced at higher latitudes.

According to the distribution of specific humidity near the surface, shown in Figure 2.31:

- Specific humidity levels throughout the tropics are about five times larger than at high latitudes.
- The air over land surfaces tends to be drier than the marine air in the winter hemisphere. In the summer hemisphere, the continental air is more moist than the marine air at the same latitude over the eastern portions of the continents, and drier in the west.

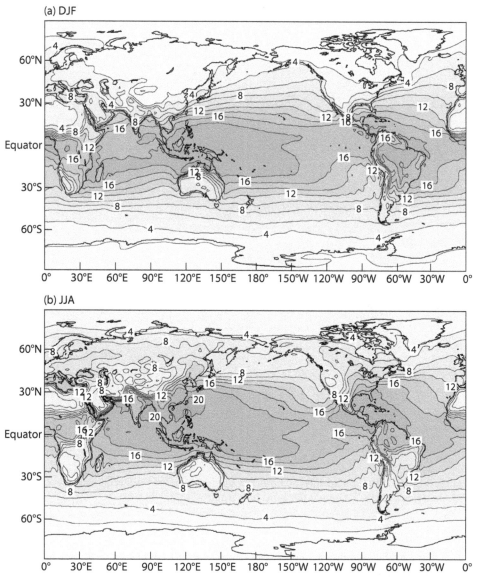

Figure 2.31 Geographical distribution of specific humidity at 900 hPa for (a) December-January-February (DJF) and (b) June-July-August (JJA). Units are 10^{-3} kg_{H_2O}/kg_{air}.

- The specific humidity distribution has features in common with the surface temperature distribution (Fig. 2.6). For example, higher specific humidity generally accompanies the higher temperatures of the western Pacific and Atlantic compared with the eastern parts of the ocean basins.
- As is the case for surface temperature, seasonality in specific humidity is greater over land surfaces.

Another measure of water vapor content is the *vapor pressure, e,* defined as the partial pressure exerted by water vapor molecules in a volume of moist air. Consider a layer of moist air overlying a plane surface of liquid water and exchanging water molecules with the surface through evaporation and condensation. For a given temperature of this atmosphere/water system, the value of the equilibrium vapor pressure, that is, the vapor pressure when the rate of evaporation from the liquid water surface is equal to the rate of condensation onto the surface, is called the *saturation vapor pressure, e_s.* Atmospheric vapor pressure is less than the saturation vapor pressure when evaporation is restricted, when the atmospheric circulation diverges moisture, or when a state of equilibrium has not been achieved. Air can also become supersaturated when *condensation* (the phase change of water from vapor to liquid phase) is somehow inhibited, for example, by the absence of condensing surfaces.

Because evaporation and condensation rates are temperature dependent, the saturation vapor pressure also depends on temperature, and that relationship is expressed by the Clausius-Clapeyron equation. For temperatures (in °C) typical of earth systems,

$$e_v(T) \cong 0.6112 \exp\left(\frac{17.67T}{T+243.5}\right), \tag{2.8}$$

as plotted in Figure 2.32. The relationship is not linear. Rather, the saturation vapor pressure becomes more sensitive to temperature under warmer conditions, for example, in the tropics. *Relative humidity, RH,* is the ratio of the vapor pressure to the saturation vapor pressure.

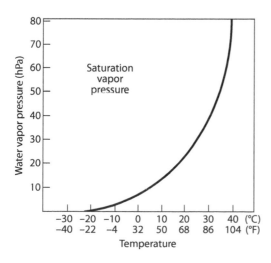

Figure 2.32 Saturation vapor pressure as a function of temperature.

Note that the physical processes determining water vapor concentrations in the atmosphere have nothing to do with the ability of air to "hold" water. This characterization of the dependence of the saturation vapor pressure on temperature is misleading.

2.4 THE CRYOSPHERE

From the Greek word *kryos*, meaning cold, *cryosphere* refers to the frozen parts of the climate system and includes sea ice, freshwater ice, ice shelves, icebergs, snow, glaciers, frozen ground, and permafrost. About 11% of the planetary surface is covered by permanent ice, and about 14% of the land surface has permanently frozen soil.

Table 2.2 provides an estimate of how frozen water is distributed within the earth system. The vast majority of the ice is sequestered on land in *glaciers*, which form in regions where the winter's snowfall does not melt completely over the summer. The minimum defined area for a glacier is 0.1 km². Ice in glaciers flows from an *accumulation area*, where the ice mass is increasing by snowfall, to an *ablation area*, where ice mass is being lost by melting or calving.

Ice sheets are large areas of glacial ice, covering 50,000 km² or more, formed over thousands of years. There are two ice sheets in today's climate, covering Antarctica and Greenland. Together, they contain more than 99% of the earth's fresh water. The Antarctic ice sheet is, by far, the largest mass of ice on the planet. Rates of snowfall onto the Antarctic continent are quite low (~150 mm/yr), so the buildup of ice is a millennial time scale process. (Note

Table 2.2 Estimates of the distribution of snow and ice in the earth system

Ice type		Mass (10^{15} kg)	Area (10^6 km²)	Flux (10^{15} kg/yr)	Residence time (yr)
Glacial ice	Antarctic ice sheet	26,400	12.0	2.2	12,000
	Greenland ice sheet	2,660	1.7	0.54	5,000
	Smaller glaciers	150	0.64	0.67	200
Ground ice		~4,000	21	~6	~700
Sea ice		35	26	33	1
Seasonal snow		10.5	72	~25	~0.4
Icebergs		7.6	64	2	4
Ice in the atmosphere		1.7	510	390	0.004

the long residence time for ice in Antarctica listed in Table 2.2.) The mass of the Greenland ice sheet is about one order of magnitude smaller than that of the Antarctic ice sheet.

More than 67,000 smaller glaciers are spread throughout the world, including the tropics high in the Andes Mountains and on African topography. Kilimanjaro Glacier, for example, is located only 3° of latitude from the equator in East Africa. Glaciers exhibit significant change on decadal time scales and longer. A glacier's *mass balance* is the difference between its accumulation and ablation rates. A positive mass balance indicates a growing glacier, and a negative mass balance indicates glacial retreat. With only a few exceptions, the world's smaller glaciers are currently in retreat as the climate warms (Fig. 2.33).

Glaciers are invaluable recorders of climate. As the fallen snow compresses and turns into ice, air bubbles are trapped within the glacier. These air bubbles are samples of the atmospheric composition at the time of ice formation. Ice cores from Vostok, in East Antarctica, are drilled to a depth of over 3.5 km and record 400,000 years of climate history. In the Northern Hemisphere, the longest cores are drilled in the Greenland ice. The ice at the bottom of these cores, at more than 3 km depth, formed more than 200,000 years ago.

In addition to providing air samples, ice cores yield information about past temperature and precipitation through isotopic analyses. Sediment deposited

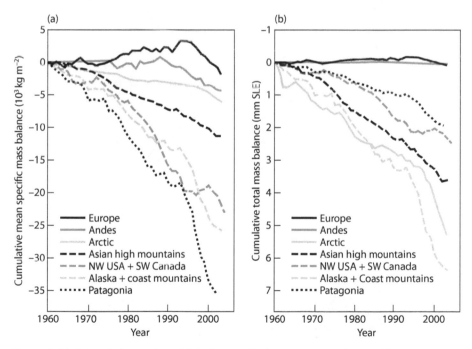

Figure 2.33 Mass balance since 1960 for small glaciers around the world. Units "mm SLE" are the number of millimeters of sea level rise (sea level equivalent) associated with the mass balance. Adapted from Kaser, G., J. G. Cogley, M. B. Dyurgerov, M. F. Meier, and A. Ohmura, 2006: Mass balance of glaciers and ice caps: Consensus estimates for 1961–2004. *Geophysical Research Letters* 33:L19501.

as ice forms records volcanic activity, deposition of extraterrestrial material and changes in the atmospheric circulation. Climate data are also extracted from tropical glaciers. For example, in the Quelccaya Ice Cap on the Peruvian Altiplano, clearly distinguished annual accumulation layers record a 1500-year chronicle of precipitation.

After glaciers, ground ice is thought to be the next largest reservoir of ice in the climate system, but estimates of mass and residence times are uncertain.

Sea ice is ice that forms in the ocean as distinguished from *icebergs*, which form on glaciers and calve into the ocean. About 15% of the ocean surface is covered by sea ice for some part of the year. Sea ice forms and grows in the winter months and melts in the summer, but some regions of the ocean are always covered with sea ice.

The top two panels in Figure 2.34 show winter and summer climatologies of sea ice extent in the Arctic for the period 1979–2000. The total area of ocean covered by sea ice in this region varies from about 16.5×10^6 km^2 in winter to about 6.5×10^6 km^2 in summer. In winter, sea ice extends into the

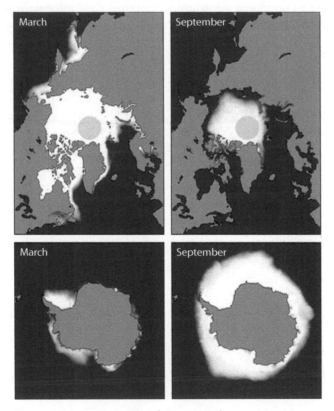

Figure 2.34 Arctic (top) and Antarctic (bottom) sea ice extent climatology for 1979 to 2000 at the approximate seasonal maximum and minimum levels based on passive microwave satellite data. Image provided by National Snow and Ice Data Center, University of Colorado, Boulder.

North Pacific and along the southern shore of the Kamchatka Peninsula, and it covers Hudson Bay and Baffin Bay. In summer, these regions are generally free of sea ice.

Antarctic sea ice extent displays an even larger seasonal variation than Arctic sea ice (bottom two panels in Fig. 2.34). This difference is, in large part, an effect of continentality. Because the southern pole is covered by land and the northern pole by water, Antarctic sea ice is located at lower latitudes than Arctic sea ice on average. Approximately 19×10^6 km^2 of the ocean around the Antarctic continent is covered with sea ice in the winter, but only about 3×10^6 km^2 survives the summer.

These large seasonal variations in polar sea ice make it clear that this component of the climate system is capable of change. Record low values of Arctic sea ice extent have been seen in recent years (Fig. 2.35a), raising concern that Arctic

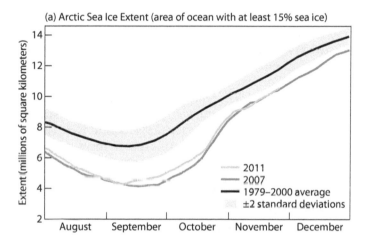

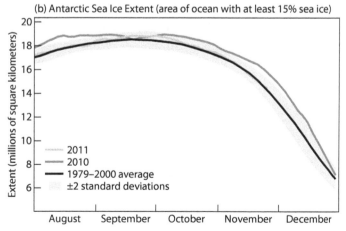

Figure 2.35 Sea ice extent in the (a) Arctic and (b) Antarctic for August through December. Image provided by National Snow and Ice Data Center, University of Colorado, Boulder (November 6, 2011).

sea ice is responding to the global warming signal. In the Antarctic, sea ice levels are maintained and even growing in association with enhanced westerly winds.

2.5 THE BIOSPHERE

Life on earth has played a crucial role in the development of the atmosphere and the physical climate system, and it plays a continuing role in maintaining and changing climate. While this aspect of climate dynamics is not a focus for this book, a few examples are provided to demonstrate interactions between the physical climate system and the planet's biology.

THE OCEAN BIOLOGICAL PUMP

One example of the influence of biological activity on climate is the role of marine life in determining the chemical composition of the atmosphere. The *ocean biological pump* is the mechanism by which CO_2 is absorbed during photosynthesis by marine organisms in the ocean's *euphotic zone*, the region into which sunlight penetrates, and then transferred deeper into the ocean's interior when the organism dies. Some of this CO_2 sequestration is temporary, and the CO_2 is returned to the ocean waters to increase the partial pressure of CO_2 when organisms decompose within the euphotic zone. But some of the CO_2 is removed from contact with the atmosphere for very long periods. For example, when phytoplankton is ingested by a clam the CO_2 can be bound into the clam's calcareous shell. When the clam dies, the shell can settle under gravity into the deep ocean. This settling of decaying phytoplankton and zooplankton, fecal pellets, shells, and various other particles is referred to as the *detritus rain*, and it transports significant amounts of CO_2 into the deep ocean.

The overall effect of the ocean biological pump is to reduce the partial pressure of dissolved CO_2 in the surface waters, allowing the additional uptake of CO_2 from the atmosphere. If the ocean biological pump were not operating, that is, with a "dead" ocean, it is estimated that the current atmospheric CO_2 concentration would be 25% greater.

EVOLUTION OF THE ATMOSPHERE

The presence of molecular oxygen (O_2) in the earth's atmosphere at 21% by volume is the result of biological activity. When photosynthesis began on earth, roughly 3.5 billion years ago, CO_2 was consumed from the atmosphere—atmospheric CO_2 levels decreased and O_2 levels increased. Very early photosynthetic activity was carried on by *prokaryotes*, which are single-cell organisms, including various bacteria species. Not all types of prokaryotes are able to undertake photosynthesis, and for those that can the process is inefficient for increasing atmospheric O_2 levels dramatically. Nevertheless, atmospheric O_2 levels increased slowly to a concentration of about 0.2% by about 2 billion years ago. Then, the prokaryotes were joined by *eukaryotes*. A eukaryote is any organism whose cells have nuclei, including both plants and animals. Photosynthesis by eukaryotic plants significantly increased O_2 production rates on the planet. But the process of building up oxygen levels in the atmosphere was still slow, because many rocks on the surface were easily oxidized and large amounts of oxygen were bound into the rocks. About 1 billion years ago these oxygen sinks on the surface became sufficiently saturated and atmospheric levels of molecular O_2 began to increase more rapidly.

Thus, while life originated on the planet in an anaerobic atmosphere, the more widespread development of biological activity over long periods of time created an oxygenated atmosphere. The continuation of biological activity is essential for maintaining O_2 in the atmosphere.

THE GAIA HYPOTHESIS

The Gaia hypothesis was developed by James Lovelock in his book *Gaia: A New Look at Life on Earth*. It suggests that the lower atmosphere of the earth is an integral, self-regulating, and necessary part of life on the planet. The idea is that, for hundreds of millions of years, life has controlled the temperature, the chemical composition, the oxidizing ability, and the acidity of the earth's atmosphere to maintain conditions to sustain, or even optimize, life.

The Gaia hypothesis is named in honor of the Greek goddess of the earth. The concept embodied by this hypothesis is that the global ecosystem actively sustains and regulates the environment for its own benefit. The idea is not that individual species act in this way, as in the Darwinian concept of natural selection, but that the biosphere as a whole acts to optimize the planet for the existence of life as a whole. The idea is not literally correct, but it is interesting and has proved useful for helping earth scientists trained in the physical sciences to consider an important role for biological activity.

The development of O_2 in the earth's atmosphere discussed previously may seem to be an example of the Gaia hypothesis at work. As life developed on the planet, CO_2 levels were drawn down and O_2 levels increased, and this made the environment more favorable for the presence of life and enabled increased biological activity.

The Gaia hypothesis ascribes a certain resiliency to life on earth, suggesting that any changes in climate are controlled to benefit biology as a whole (but not necessarily individual species). Currently, human activity is changing climate. Is this process optimizing the environment to benefit life on the planet? That seems doubtful when we consider the ongoing mass extinction due to human activity and the threat to ecosystems in rapidly shifting climate zones.

2.6   DATA SOURCES AND REFERENCES

The National Snow and Ice Data Center maintains an excellent website for general education on the cryosphere as well as more in-depth exploration of recent findings.

Figures of atmospheric variables in this chapter, and subsequent chapters, use the NCEP/NCAR and ERA-Interim reanalyses described in the following references:

Kistler, R., E. Kalnay, W. Collins, and co-authors, 2001: The NCEP-NCAR 50-year Reanalysis: Monthly Means CD-ROM and Documentation. Bulletin of the American Meteorology Society, 82, 247–268.

Dee, D. P., S. M. Uppala, A. J. Simmons, and co-authors, 2011. The ERA-Interim Reanalysis: Configuration and performance of the data assimilation system. Quarterly Journal of the Royal Meteorological Society, 137, 553–597.

Observations of the ocean are plotted from the World Ocean Atlas:

Conkright, M. E., S. Levitus, T. O'Brien, and co-authors, 1998: World Ocean Database (1998) Documentation and Quality Control. National Oceanographic Data Center, Silver Spring, MD.

2.7 EXERCISES

2.1. Calculate the following values:

(a) The lapse rate in the troposphere in middle latitudes (say, between 40°N and 60°N) of the Northern Hemisphere.

(b) The zonal mean meridional temperature gradient between 40°N and 60°N near the surface of the earth.

Express your answers in K/km, and note the relative size of the two temperature gradients.

2.2 The atmosphere is thin.

(a) A typical horizontal space scale for weather systems in midlatitudes is 1000 km. Assuming that weather systems occupy the full depth of the troposphere, what is the ratio of the horizontal and vertical space scales typical of these systems?

(b) How does the magnitude of the meridional temperature gradient in the atmosphere compare with the magnitude of the lapse rate?

2.3 Wind-driven ocean currents.

Compare the magnitudes and directions of ocean surface currents with the magnitudes and directions of surface winds. Generalize. Roughly, by what factor do the low-level wind and the surface current magnitudes differ? Where do their directions generally agree, and where do they differ?

2.4 Plot vertical profiles of specific humidity at the equator and at 50°N using pressure as a vertical coordinate.

2.5 On average for the globe, what is the ratio of the mean evaporation rate from a unit surface area of ocean to the mean evaporation rate from a unit surface area of land? What, do you think, is the most important reason for the difference? (Note: The surface of the earth is about 71% ocean.)

3

OBSERVATIONS OF NATURAL CLIMATE VARIABILITY

Chapter 2 provided a description of the time-mean state of the climate system. But to characterize this complex system more completely requires considering more than the climatology, since there is pronounced variability in the system on a wide range of time and space scales. This is especially important if we are concerned about isolating and understanding the human-induced climate change signal. To differentiate climate change from climate variability, we need to understand the features, processes, and causes of both.

There are two types of climate variability—*internally generated* and *externally forced*. Internally generated variability results from processes within a system, while externally forced variability arises when some factor outside the system causes change.

Classification of internally generated and externally forced variability depends on how the boundaries of a system are defined. For example, if we were to study the atmosphere in isolation from the rest of the climate system, then changes in sea surface temperatures would be termed external forcing. But in the context of the coupled ocean/atmosphere system, variability induced in the atmosphere by variations in sea surface temperatures are internally generated variability. In the case of external forcing, changes within the system do not feed back to modify the forcing agent. One example of external *climate forcing* is climate change caused by variations in the amount and/or distribution of solar energy incident on the earth due to changes in the solar luminosity or the earth's orbital parameters.

Climate variations can also be classified according to their characteristic space and time scales. Typical classifications by space scale are *local*, *regional*, *continental*, and *global*, corresponding roughly to 1–50 km, 50–1000 km, 1000–10,000 km, and 10,000–40,000 km, respectively. Time scales of variability include *diurnal*, *intraseasonal*, *seasonal* or *annual*, *interannual*, *decadal*, and *millennial*. Any of these variability signatures may be internally generated or externally forced. This approach to classifying variability is useful because the causes and processes of climate variability in the climate system vary with time and space scales.

3.1 DIURNAL AND SEASONAL CLIMATE VARIATIONS

Table 3.1 provides estimates of ranges in atmospheric temperature on diurnal and annual time scales. Note that diurnal temperature ranges have the following characteristics:

- They are greater on continents than over the ocean at all latitudes. This is related to differences in the heat capacities of land and water surfaces (chapter 5).
- They are larger in the subtropics than in the tropics. The subtropical atmosphere has lower specific humidity than the tropical atmosphere (Fig. 2.31). Water vapor is a powerful greenhouse gas and, in its absence, strong radiative cooling to space decreases nighttime temperatures and increases the diurnal temperature range (chapter 4).
- They are very small over the oceans. Because of its high heat capacity, the ocean surface barely reacts to the relatively high frequency diurnal forcing (chapter 5).

Table 3.1 Approximate temperature ranges on diurnal and seasonal time scales

Latitude	Diurnal		Seasonal	
	Continental	Marine	Continental	Marine
Southern Hemisphere high latitudes 90°S–60°S	8 K	2 K	30 K	12 K
Southern Hemisphere middle latitudes 60°S–35°S	5 K	3 K	8 K	5 K
Southern Hemisphere subtropics 35°S–15°S	15 K	2 K	15 K	3 K
Tropics 15°S–15°N	10 K	2 K	3 K	3 K
Northern Hemisphere subtropics 15°N–35°N	17 K	3 K	16 K	4 K
Northern Hemisphere middle latitudes 35°N–60°N	13 K	4 K	35 K	8 K
Northern Hemisphere high latitudes 60°N–90°N	6 K	3 K	33 K	20 K

Note that seasonal temperature ranges have the following characteristics:

- They are larger than diurnal temperature ranges in high and middle latitudes, comparable in the subtropics, and smaller in the tropics.
- They are larger on continents than over the oceans at all latitudes, as is the case for diurnal temperature ranges.
- They are larger in Northern Hemisphere middle latitudes than in Southern Hemisphere middle latitudes. This is a continentality effect, due to the greater extent of continents in the Northern Hemisphere.

Seasonality in the tropics and much of the subtropics is more sharply defined by precipitation than by temperature in most regions. Over India, West Africa, and central South America, precipitation rates are about an order of magnitude greater in the summer months than in the winter months (Figure 3.1). In other regions, for example, over parts of East Africa, there are two rainy seasons during the transition seasons (spring and fall) with drier conditions in the solstitial seasons (summer and winter).

The seasonal cycle of precipitation is less pronounced over the oceans than over continents. This is particularly true across the tropical North Pacific Ocean where, as seen in Figure 2.27, the ITCZ precipitation maximum does not follow the sun into the Southern Hemisphere during the Northern Hemisphere winter months.

A primary reason for the strong seasonality in precipitation at lower latitudes is the occurrence of monsoons. The traditional definition of a monsoon climate is one in which the direction of the low-level wind reverses by 180° between summer and winter, although this definition is often not strictly applied. Figure 2.13 shows this reversal over India, as an example. In the summer months, the flow is directed onto the west coast from the Arabian Sea as the Somali jet carries moisture across the equator to feed powerful rainfall systems. In the winter, however, the flow is directed off the continent during the dry "winter monsoon." The physics of monsoon circulations is discussed in chapter 7.

The strongest monsoon system on the planet is the Asian monsoon, which covers much of southern Asia in boreal summer. Ninety percent of the annual rainfall in Mumbai, on the west coast of India, is delivered during the monsoon season and in Chennai, on the east coast, 60% of the annual precipitation is monsoonal. Other regions with monsoon climates are West Africa, Australia, South America, and the southwestern United States.

3.2 INTRASEASONAL CLIMATE VARIABILITY

It is easy to understand how diurnal and seasonal variability are generated in the climate system since insolation provides external forcing on these time scales. But the climate system also varies on longer time scales. Climate variations on time scales less than 3 months but greater than about one week are known as intraseasonal variations. One prominent mode of climate variations on this time scale is the *Madden-Julian oscillation* (MJO). The MJO is a tropical oscillation that is apparent in precipitation, cloud, temperature, geopotential height, and circulation observations throughout the full depth of the troposphere with periods between 40 and 60 days.

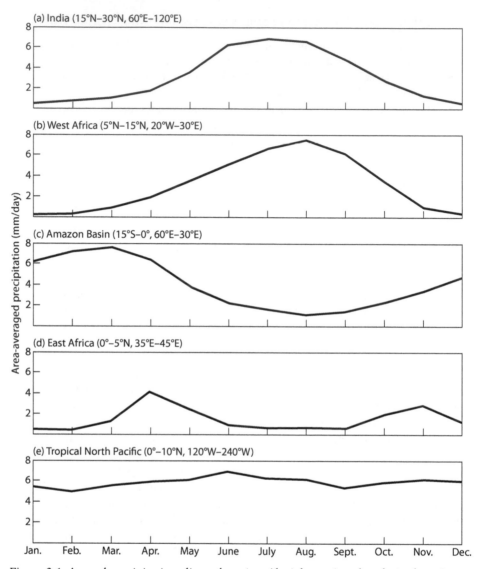

Figure 3.1 Annual precipitation climatology (mm/day) for various low-latitude regions.

To form a *composite,* the phases of numerous MJO observations are aligned to define a typical anomaly (Fig. 3.2). The MJO begins with the development of strong convection and positive precipitation anomalies in the western Indian Ocean. The anomalies are associated with low surface pressures, converging low-level winds, and diverging upper-level winds (top panel). At the same time, the western Pacific is dry. About 2 weeks later (second panel), the positive precipitation anomaly has moved to the eastern Indian Ocean, and 2 weeks after that (third panel) the western Pacific receives anomalously high rainfall. The wet and dry anomalies continue to propagate eastward at about 5–10 m/s, dissipating in the central Pacific where the surface waters are cooler. In some cases the oscillation is reinvigorated and becomes apparent again in the tropical

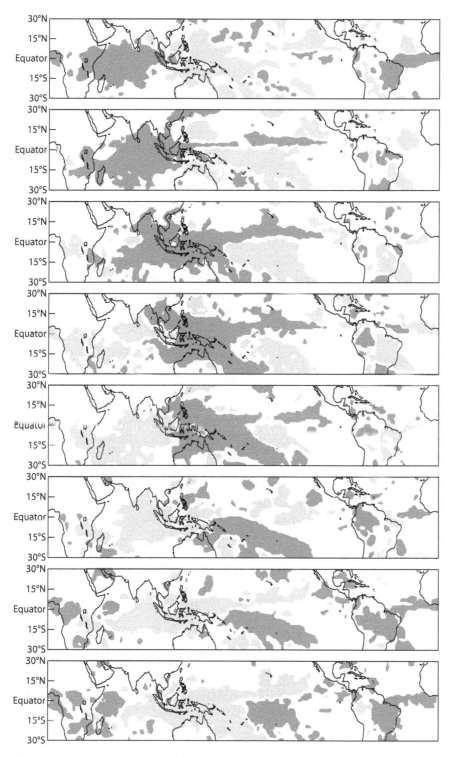

Figure 3.2. Composites of precipitation anomalies for eight periods within the MJO. Positive rainfall anomalies are indicated by dark shading, and negative anomalies by lighter shading. Modified from www.cpc.ncep.noaa.gov.

Atlantic. The MJO is superimposed on the monsoon circulations in Asia, Australia, Northern America, Africa, and South America and can lead to intraseasonal variations in those systems. It can also influence the location and strength of both the SACZ and the SPCZ, and the distribution of tropical cyclones.

3.3 INTERANNUAL CLIMATE VARIABILITY

Climate variations also occur on time scales of a few years, called interannual variability. The most prominent mode of interannual variability in the climate system is a coupled atmosphere/ocean oscillation known as the *El Niño— Southern Oscillation,* or *ENSO.*

Figure 3.3 is a *Hovmöller diagram* depicting the time evolution of sea surface temperatures in the equatorial Pacific from January 1995 through January 2003. The western Pacific (~120°E to 180°) is characterized by relatively constant, warm temperatures without evidence of a seasonal cycle. In the eastern Pacific, east of 140°W, the seasonal signal has an amplitude of about 5 K, with coolest temperatures in early boreal spring and warmest temperatures in late fall. This seasonal oscillation was interrupted in 1997, however, as warm waters stretched across the entire tropical Pacific. This type of episode is known as an *El Niño event*, or an *ENSO warm event.*

A map of sea surface temperature for DJF in 1997/98 is shown in Figure 3.4a. Comparison with the climatological sea surface temperature distribution (Fig. 2.16a) shows that an ENSO warm event is characterized by a relaxation of the longitudinal temperature gradient across the tropical Pacific Ocean. The sea surface temperature anomaly (Fig. 3.4b) is centered on the equator, exceeding 4 K in the eastern Pacific. Note that there are no sea surface temperature anomalies in the western Pacific during this warm event.

Returning to Figure 3.3, we see that the cool waters of the eastern Pacific extended farther west than usual in late 1998 and 1999. (Track the position of the 300 K isotherm.) These occurrences are called *La Niña events*. A comparison of the climatological sea surface temperature distribution for DJF (Fig. 2.16a) with the distribution during a strong La Niña event in 1998/99 shown in Figure 3.5a indicates that during a cool event the eastern Pacific cool tongue is exceptionally well developed and extends farther west along the equator. The La Niña sea surface temperature anomaly, shown in Figure 3.5b, is typically weaker in magnitude than the El Niño anomaly (Fig. 3.4b) and located farther west.

Sea surface temperature anomalies in the area denoted by the rectangles in Figures 3.4b and 3.5b are often used to monitor ENSO events. This is the *Niño 3.4 region.* (It was formed by a refinement of two other monitoring areas known as the *Niño* 3 and *Niño* 4 regions.) Figure 3.6a shows the record of Niño 3.4 sea surface temperature anomalies (the deviations from the 1971–2000 mean temperature) from 1900 through 2009. The record has been smoothed using an 11-month running mean to emphasize variations on interannual time scales. The telltale warming of an El Niño is evident in many fall and winter seasons, including strong events in 1941/42, 1972/73, 1982/83, and 1997/98. Strong cool events occurred in 1955/56, 1975/76, 1988/89, and 2000/2001.

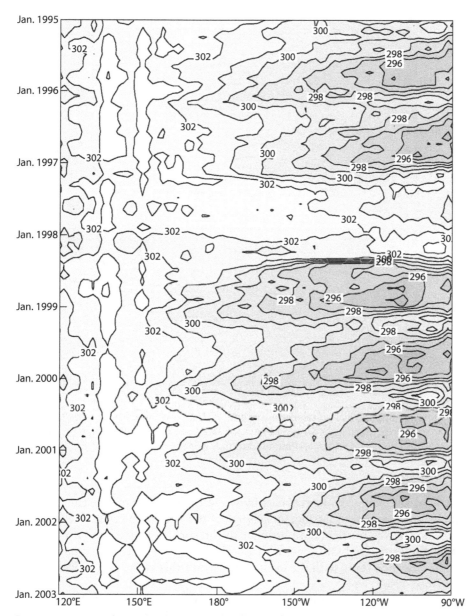

Figure 3.3 Sea surface temperature across the Pacific Ocean on the equator plotted as a function of time (y-axis) in a Hovmöller diagram.

But these sea surface temperature anomalies are only one aspect of the ENSO oscillation. As the warm waters spread eastward across the equatorial Pacific during a warm phase, the thermocline depth in the eastern Pacific (Fig. 2.18a) deepens and sea level rises. The low precipitation rates of the central and eastern Pacific (Fig. 2.26) are replaced by higher rainfall rates as deep convective cloudiness shifts eastward from the western Pacific warm pool. At the same time, the low-level easterly winds over the eastern Pacific (Fig. 2.13)

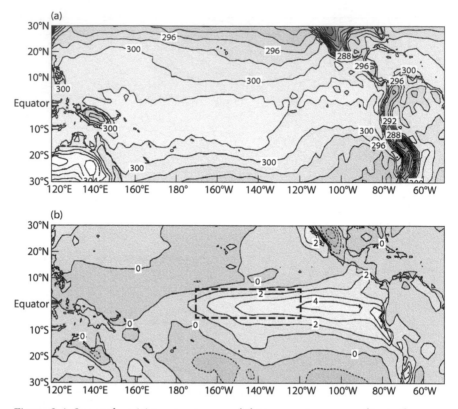

Figure 3.4 Sea surface (a) temperatures and (b) temperature anomalies in the tropical Pacific during the strong El Niño of 1997/98. The rectangle in (b) marks the Niño3.4 region, which is often used to monitor the progression of El Niño and La Niña events.

weaken, and may even be replaced by westerly flow. When the easterly winds slacken in the eastern Pacific, upwelling off the South American coast diminishes and fewer nutrients are brought to the surface to feed aquatic life. Because the associated failure in fishing typically occurs near Christmas, people connected the events with the birth of Jesus and this is the origin of the name El Niño.

These changes in atmospheric circulation systems are recorded in sea level pressures across the Pacific. An El Niño event is characterized by a decrease in the climatological high pressure (high geopotential heights) in the eastern Pacific and an increase in geopotential heights in the western Pacific. In the early 1900s Gilbert Walker, a British scientist on assignment as the head of the Indian Meteorological Service, recorded the inverse correlation between surface pressure in the western Pacific, represented by Darwin, Australia, and the central Pacific, represented by Tahiti, which he called the *Southern Oscillation*. Smoothed sea level pressure at these two locations is plotted in Figure 3.6b, and it is clear that when surface pressure is anomalously high in the western Pacific it is anomalously low in the central Pacific, and vice versa. During particularly strong ENSO warm events (e.g., 1982/83 and 1997/98), pressures at Darwin and Tahiti may be equal.

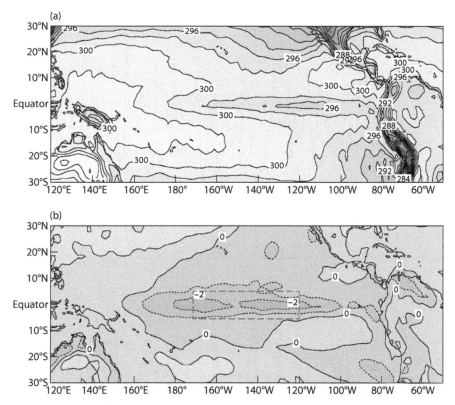

Figure 3.5 Sea surface (a) temperatures and (b) temperature anomalies in the tropical Pacific during the ENSO cool event of 1998/99. The rectangle in (b) marks the Niño3.4 region, which is often used to monitor the progression of El Niño and La Niña events.

The *Southern Oscillation Index* (SOI) is a metric commonly used to track and quantify the ENSO oscillation. It is based on the surface pressure anomalies at Tahiti minus those at Darwin, so negative values of the SOI indicate an ENSO warm event. Figure 3.6c shows the time series of the SOI. The periods 1982/83 and 1997/98 stand out as particularly strong warm events, and a series of weak warm events occurred through the early 1990s. Other indexes may be used to monitor ENSO, for example, the *Multivariate ENSO Index*, which takes into account surface wind, surface temperature, and cloudiness anomalies in addition to surface pressure.

The *quasi-biennial oscillation* (QBO) is an example of interannual variability in the stratosphere; it is a result of interactions between the stratosphere and the troposphere in the tropics. The oscillation occurs in the middle and lower tropical stratosphere, where the wind direction shifts from easterly (westward) to westerly (eastward) every 26 months or so.

Figure 2.10 suggests that the climatology of the zonal mean zonal wind in the lower tropical stratosphere is small and easterly when averaged over many years. But this is misleading because in averaging over a number of years, strong positive (or westerly) wind speeds are averaged with strong negative (or easterly) wind speeds to produce apparently small mean velocities in the climatology.

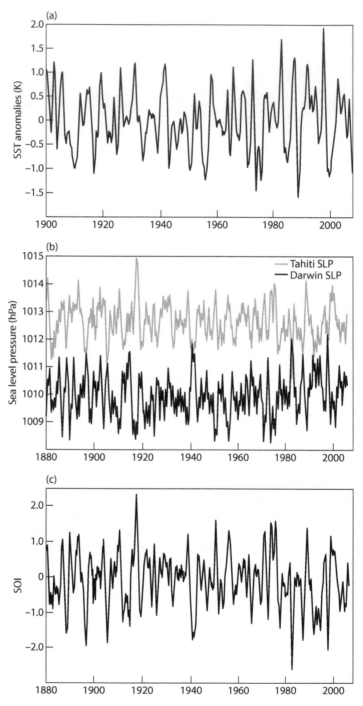

Figure 3.6 (a) Niño 3.4 sea surface temperature anomalies (difference from the 1971–2000 mean), smoothed; (b) Sea level pressure (SLP) at Darwin (black) and Tahiti (gray), 11-month running mean; (c) Southern Oscillation Index (SOI), 11-month running mean. From NOAA's Climate Diagnostic Center.

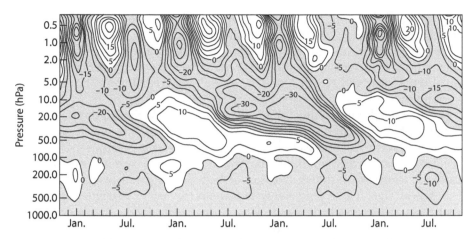

Figure 3.7 Monthly mean zonal mean wind (m/s) at 1.25°N for November 1991–November 1995. Easterly flow is shaded, and the contour interval is 5m/s.

Figure 3.7 shows the zonal wind speed as a function of time for 4 years through the depth of the tropical stratosphere at 1.25°N. Above about 4 hPa, in the upper stratosphere (see Fig. 2.8), the wind direction changes sign every 6 months. This is a seasonal signal, forced by insolation changes, called the *semiannual oscillation.*

Much longer periodicity is evident below 4–5 hPa, where zonally uniform easterly and westerly phases alternate every 24–30 months (26 months is the average). This is the QBO. Successive phases of the QBO start above 10 hPa and propagate downward, reaching the lower stratosphere about 26 months later. Meanwhile, another phase of the QBO, with winds in the opposite direction, has begun in the upper stratosphere. As the signal propagates downward, there is no loss in magnitude until the lower stratosphere where the wind speed is strongly damped.

The QBO is thought to be caused by atmospheric wave breaking. The waves are generated by convection in the upper tropical tropopause. They travel upward through the stratosphere until conditions there cause them to break—much like waves in the ocean break at the shore. On breaking, the waves deposit energy and accelerate the flow.

3.4 DECADAL CLIMATE VARIABILITY

Although this may not seem like a very natural time scale for the climate system, there is significant variability on decadal time scales. One example is "inter-ENSO" variability, that is, the fact that one ENSO can be different from another, or that there may be decades in which there are more ENSO events than in others (see Fig. 3.6). It is particularly important to understand natural variability on this time scale because it is also the time scale of human-induced changes in chemical composition of the atmosphere (see chapter 10).

The *North Atlantic Oscillation* (NAO) is a prominent mode of oscillation in the climate system that results in inverse correlations among variables over the northern and subtropical Atlantic on decadal time scales. Coherent fluctuations associated with the NAO are detectable over northern Europe, northern Africa, and North America, where they contribute to decadal-scale variability.

In the boreal winter climatology, there is a low off the southern coasts of Greenland and Iceland in the North Atlantic (see Fig. 3.8, which is a close-up of the North Atlantic section of Fig. 2.4a with finer contours). This is the *Icelandic low*. Farther south, off the Atlantic coast of Spain and northern Africa, the climatology shows a region of high geopotential heights known as the *Azores high*. These pressure systems tend to vary in strength together, so that when the Azores high is especially strong the Icelandic low is deep. That covariation is called the NAO and is an example of a *teleconnection* that links different parts of the climate system over large space scales. Because surface pressure is a measure of the weight of the overlying air (Eq. 2.1), the NAO can

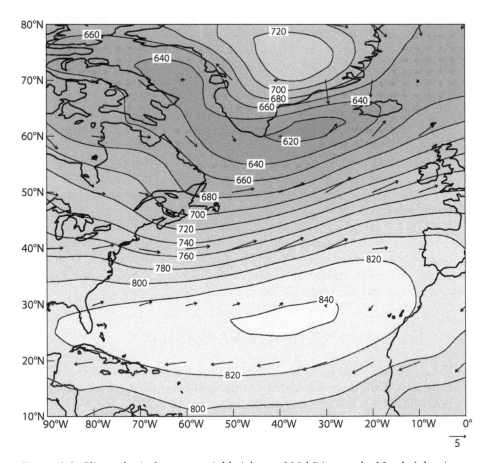

Figure 3.8 Climatological geopotential heights at 900 hPA over the North Atlantic.

be described as a meridional shifting of the atmospheric mass, north and south across the North Atlantic Ocean.

An index used to measure the strength and phase of the NAO is similar to the SOI in that it is based on surface pressure differences between two locations. In contrast to the SOI, which uses locations at similar latitudes but different longitudes, the NAO index is based on stations at similar longitudes but different latitudes. The northern location is Stykkisholmur, Iceland (65°N, 23°W). For the southern location, different studies have used Ponta Delgada (Azores), Lisbon, and Gibraltar. The NAO index is positive when both the Icelandic low and the North Atlantic subtropical high are stronger than normal; a negative NAO index indicates that both are weaker than normal. The positive phase of the NAO is often associated with warm winters over the eastern United States, warm and wet winters over northern Europe, and dry winters over southern Europe as the North Atlantic storm track (Fig. 2.14) shifts to the north.

Figure 3.9a shows the NAO index for 1950–2011 and the propensity for one phase or another to dominate on decadal time scales. For example, the negative phase of the NAO occurred more frequently than the positive phase through the 1950s and 1960s, but the positive phase was dominant from the mid-1980s until the mid-1990s. Recently, the NAO has become more neutral, and even negative.

The NAO can be seen as part of a larger pattern of variability in the Northern Hemisphere known as the *Arctic Oscillation* (AO). The negative phase of the AO is associated with anomalously high pressure over polar regions and anomalously low pressures in middle latitudes, which leads to cold air intrusions into western Europe and the central United States. The positive phase of the AO index (Fig. 3.9b) is associated with anomalously low pressures over the Arctic, with wetter conditions in northern Europe and drying in Spain, northern Africa, and the Middle East. Like the NAO index, the AO index was positive during the later 1980s and early 1990s but fairly low and variable in the 2000s.

Another example of decadal variability in the climate system is the *Pacific Decadal Oscillation* (PDO). PDO anomalies are similar in structure to ENSO anomalies except they are much longer lived and have smaller magnitudes (~0.5 K). Values for more than 100 years of a PDO index are shown Figure 3.10. A positive, La Niña–like phase of the PDO occurred from 1890–1924, characterized by anomalously warm sea surface temperatures in the western Pacific (especially at middle and high latitudes) and cool temperatures in the eastern Pacific. A warm PDO phase persisted from 1925 until 1946 and was replaced by another cool phase from 1947 until 1976. The warm phase that followed completed two cycles of the PDO in about a century.

Warm phases of the PDO are correlated with increased marine ecosystem activity and productivity in Alaska and reduced activity off the west coast of the United States farther south. This pattern is reversed during the cool phase of the PDO. The causes of the PDO are not completely understood, but recognizing its existence aids prediction because of the long persistence of the pattern.

Another type of decadal variability within the climate system occurs in the form of persistent drought. Over the central United States, the Dust Bowl

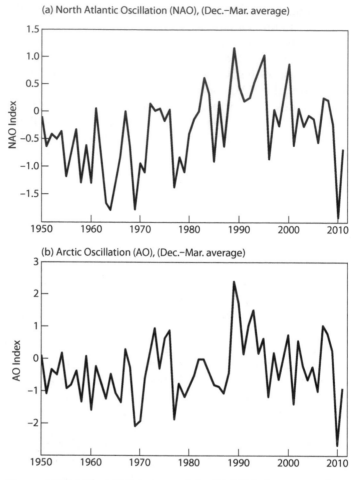

Figure 3.9 (a) The NAO index and the (b) AO index averaged over the Northern Hemisphere winter months (December, January, February, and March).

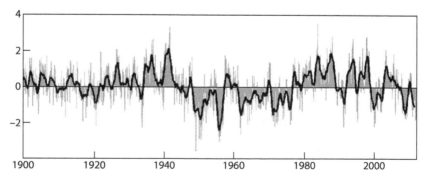

Figure 3.10 Monthly values of the PDO index (gray lines) with the smoothed time series in black. Obtained with permission from the University of Washington's Joint Institute for the Study of the Atmosphere and Oceans (http://jisao.washington.edu/pdo/).

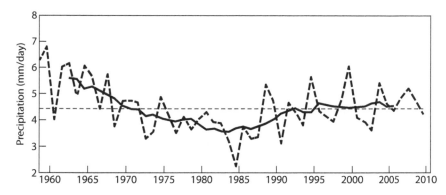

Figure 3.11 The July through September precipitation (mm/day) averaged over the western and central Sahel (10°W to 10°E and 11°N to 16°N) is shown by the dashed line; the solid line is a smoothed version to show decadal variations. Figure courtesy of Bing Pu.

climate of the 1930s combined with economic depression forced a massive westward migration, and persistent warm, dry conditions returned in the 1950s and again in the late 1980s. On longer time scales, tree ring data analysis reveals the occurrence of multidecadal *megadroughts* in past centuries.

Northern Africa is also known to be susceptible to decadal-scale drought. Recent examples are the Sahelian drought of the 1970s–1980s and the recent drought over Sudan at the beginning of the twenty-first century. Figure 3.11 shows the summer rainfall record from the Sahel (11°N–16°N, 10°W–10°E) for 1950–2010. Values plotted are rainfall differences from the mean climatology. It is clear that Sahel rainfall has significant decadal-scale variability. Dry conditions persisted through the 1970s and 1980s, for example, following two decades of relatively wet conditions in the early 1950s.

3.5 CLIMATE VARIATIONS ON CENTURY TO BILLION-YEAR TIME SCALES

In addition to the relatively short time scales already reviewed, climate variations occur on time scales of centuries and millennia, and even millions and billions of years. The study of earth's *paleoclimates* informs the study of present and future climate by demonstrating that the climate system is capable of great change, for example, from a warm planet some 70 million years ago that was hospitable to dinosaurs to the ice age world of only 20,000 years ago. And it is clear from the paleoclimate record that climate change can occur abruptly and lead to mass extinctions of flora and fauna.

The study of paleoclimates is founded on the availability of *proxy data* derived from fossil pollen, ocean and lake sediment cores, tree rings, ice cores, and coral. Various historical records such as diaries, agricultural logs, shipping records, and newspaper articles are also used to reconstruct past climate.

One well-known example of climate variation on the centennial time scale is the *Little Ice Age*. Average global temperatures are estimated to have been approximately 1.5 K cooler than today from about 1560 to1850 due to large amounts of volcanic activity, with significant stratospheric injection of aerosols, and reduced solar activity (chapter 4). Cooling was not sufficient for ice sheets to grow, but more sea ice formed in the North Atlantic and mountain glaciers in Europe advanced. Long winters shortened the growing season by up to 2 months. The resulting malnutrition accompanied by wet summers favored the spread of disease such as the bubonic plague, which killed more than one-third of Europeans.

Changes in the solar constant on millennial time scales force the *glacial/interglacial oscillation*, that is, the variation of global climate between colder and warmer climate states that is characteristic of the last 2.6 million years (the *Quaternary* geologic period). These variations in insolation are attributed to regular, predictable changes in the earth's orbital parameters according to the "astronomical theory of climate change," or the *Milankovich theory* in honor of the Serbian mathematician who developed this idea. The earth's orbit about the sun is perturbed by the gravitational pull of other planets, especially Jupiter and Saturn.

The following three attributes of the earth's orbit around the sun change with time:

- The eccentricity, e, of the earth's orbit changes with a 100,000-year periodicity, from a nearly circular orbit ($e = 0.005$) to a more elongated orbit ($e = 0.061$). Today's value is $e = 0.012$, a relatively small value.
- The tilt between the earth's axis of rotation and the plane of its orbit varies with a period of 41,000 years. Today it is 23.5°, but it ranges between 21.8° and 24.4°.
- The precession of the equinox changes the season during which the earth is closest to the sun (perihelion) over a 23,000-year cycle. Today, perihelion occurs during Northern Hemisphere winter, while 11,500 years ago perihelion occurred during Northern Hemisphere summer. In today's orbital configuration, the earth system intercepts 6% more solar energy in January than in July. When the earth's orbit is at its maximum eccentricity, this difference is as large as 30%.

These changes in the amount and distribution of solar energy fueling the earth system influence climate. For example, low values of the tilt favor cooler summers in the Northern Hemisphere. Because most of the land mass has been located in the Northern Hemisphere throughout the Quaternary, low tilt is associated with ice age climates—continental ice sheets must survive the summer months if glaciers are to grow.

The periodicities with which the orbital parameters change (23,000, 41,000, and 100,000 years) are clearly found in the geologic record, but the relationships between the orbital parameters and climate characteristics are not simple or synchronous. For example, the peak of the last ice age was about 21,000 years ago when the orbital parameters were similar to those of today.

On even longer time scales, climate change forcing factors include the distribution of the continents, orographic uplift, the sun's evolution (chapter 4 has an example), and the evolution of the atmosphere's chemical composition (section 2.5). The difference between the present-day globally averaged surface air temperature and the Mesozoic temperature (the Age of the Dinosaurs) is about 15 K (15°C/27°F). The glacial–interglacial oscillation of climate is a relatively recent feature of the climate system and is thought to have occurred only during the Quaternary (the last 1.8 million years). The amplitude of this oscillation, that is, the difference in the global temperature between a glacial and an interglacial climate, is less than 10 K (10°C/18°F).

It is clear from modern observations, historical records, and paleoclimate reconstructions that the earth's climate changes on all time scales and that these changes are of sufficient magnitude to influence life on the planet. All these changes have a cause and an explanation, whether related to some external forcing or to internal workings of the climate system, but they may not be fully explained yet. The key to understanding climate variability—and climate change—is to study the physics and dynamics of the climate system. We begin that study in the next chapter with a consideration of the energy source of the climate system—solar radiation.

3.6 ADDITIONAL READING

A number of excellent sources of information are maintained on the Web to provide background as well as the latest observations of various modes of climate variability. Examples include the websites of NOAA's Climate Prediction Center and NASA's Earth Observatory.

4

RADIATIVE PROCESSES
IN THE CLIMATE SYSTEM

The absorption of solar radiation is the primary energy source of the earth system. Other sources are geothermal energy, tidal energy, and waste heat from the burning of fossil fuels, which together contribute less than 0.03% to the total energy flux into the climate system. So it is appropriate that our discussion of the physical processes important for climate dynamics starts with a consideration of radiative processes. We begin with a review of some basic physical principles.

4.1 BLACKBODY THEORY

A *blackbody* is a theoretical object that absorbs all radiation incident onto its surface and reradiates all that energy. The total amount of energy emitted from a unit area of a blackbody each second (i.e., the power) is given by the Stefan-Boltzmann law:

$$E = \sigma T^4,$$ (4.1)

where $\sigma = 5.67 \times 10^{-8}$ J/(m$^2 \cdot$ s $\cdot$ K^4) is the Stefan-Boltzmann constant. In Eq. 4.1, T is the temperature of the blackbody, which is assumed to be isothermal. This relationship was first developed experimentally (by Josef Stefan) and later derived theoretically (by Ludwig Boltzmann).

Some materials in the real world obey the Stefan-Boltzmann law quite closely, while others do not. Thermal emission from these materials is generally proportional to the fourth power of their temperature, but the amount of energy emitted is generally lower than that of a perfect blackbody. A better approximation for the energy radiated by real materials results by introducing the blackbody *emissivity*, ε, defined as the fraction of the incident energy that is reemitted. (The dependence of emissivity on wavelength is neglected here.) If we take emissivity into account, the energy emitted by a *"gray body"* is

$$E = \varepsilon \sigma T^4.$$ (4.2)

Emissivity values for substances relevant to climate are generally close to 1 (see chapter 5 and Table 5.1).

Blackbody radiation emission is distributed across wavelength, λ, according to the Planck formula:

$$B_\lambda(T) = \frac{2hc^2}{\lambda^5 [\exp(hc/\lambda kT) - 1]},$$ (4.3)

where $B_\lambda(T)$ is the *spectral radiance*, defined as the rate at which energy is emitted per square meter per unit wavelength per steradian (the unit of solid angle). The units of $B_\lambda(T)$ are W/(m$^2 \cdot$ sr $\cdot \mu$m) when wavelength is in microns (1 micron = 1 μm = 10^{-6} m). In Eq. 4.3, h is Planck's constant (6.63×10^{-34} J·s), c is the speed of light (3.00×10^8 m/s), and k is Boltzmann's constant (1.38×10^{-23} J/K).

Note the following attributes of blackbody radiation:

- For a blackbody with temperature T, σT^4 is the rate of energy emission per unit area (W/m^2), integrated over all wavelengths and over the hemisphere into which the surface radiates: [1]

$$\sigma T^4 = \pi \int_0^\infty B_\lambda d\lambda. \qquad (4.4)$$

- As the temperature of the emitting body increases, the wavelength of maximum emission, λ_{MAX}, decreases according to Wein's displacement law:

$$\lambda_{MAX} = \frac{2.898 \times 10^{-3}}{T}, \qquad (4.5)$$

  where T is in Kelvin and λ_{MAX} is in μm.
- According to Kirchhoff's law, the emissivity, ε, of a blackbody at any given wavelength is equal to its *absorptivity*, a, for that same wavelength. A strong absorber is an equally strong emitter. Kirchhoff's law holds for any object at a constant temperature.

For a given temperature, Eq. 4.3 can be used to calculate energy emitted as a function of wavelength. The resulting plots are called *Planck curves*.

4.2 APPLICATION OF BLACKBODY THEORY TO THE EARTH SYSTEM

Imagine the earth as a spherical blackbody in empty space. Now, switch on the sun and let solar radiation fall onto the surface of the sphere. Because it is black, the sphere will absorb all the incident solar radiation. It will begin to heat up and emit energy in the form of radiation at a rate, E in W/m^2, that is dependent on its temperature, T, according to the Stefan-Boltzmann law (Eq. 4.1). The temperature of this model earth will continue to increase until the rate at which it is emitting energy is equal to the rate at which it is absorbing energy. This final state is called *radiative equilibrium*, and the temperature at which this occurs is called the *radiative equilibrium temperature*, T_E.

For this idealized example and the calculations that follow, consider the long-time average over the diurnal and seasonal cycles and over many years. This is equivalent to imagining that the energy incident on the earth is distributed uniformly over the entire sphere, and isothermal conditions prevail.

---

[1] The details of this integration are not presented. The reader is referred to textbooks on radiation for guidance on the integration by parts.

Under these simple assumptions, T_E for the earth can be calculated if the rate at which energy is absorbed by the system is known, since in a state of radiative equilibrium the energy emitted is equal to the energy absorbed. If S_{ABS} denotes the rate at which solar radiation is absorbed by a unit surface area of the black spherical earth model, then

$$\sigma T_E^4 = S_{ABS} \Rightarrow T_E = \left(\frac{S_{ABS}}{\sigma}\right)^{1/4}. \tag{4.6}$$

To calculate an expression for S_{ABS} we first consider the average rate at which solar energy is incident on a unit surface area of the earth, S_{INC}. The rate at which solar energy is incident on a unit area perpendicular to the solar beam at the earth–sun distance (drawn in Fig. 4.1a) is known as the *solar constant*, S_0. (The inconstancy of the solar "constant" is discussed below.) It is equal to the total energy output of the sun, or the solar luminosity, L_S, divided by the area over which that energy is distributed:

$$S_0 = \frac{L_S}{4\pi r^2} = \frac{3.9 \times 10^{26}\,\text{W/m}^2}{4\pi (1.5 \times 10^{11}\text{m})^2} = 1380\,\text{W}, \tag{4.7}$$

where r is the average earth–sun distance. This simple calculation yields a value for the solar constant that is a little high, since reduction of the solar beam by interactions with interplanetary dust is not taken into account. A more representative value is $1368\,\text{W/m}^2$, although S_0 varies by a few W/m^2 (see the following section).

S_0 is greater than S_{INC} because the only location that receives insolation at the value of S_0 is the subsolar point. In calculating S_{INC} we need to account for

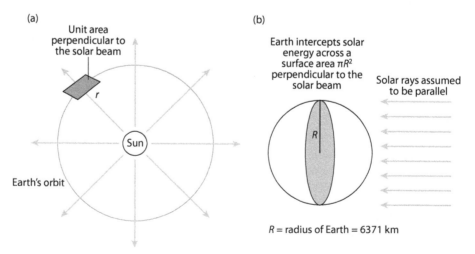

Figure 4.1 Calculation of the average solar radiation incident on a unit surface area of the earth.

the fact that half the earth is receiving no solar energy at any given time, and also for the oblique angle of incidence of the solar rays away from the subsolar point. Figure 4.2b shows that the earth intercepts energy from the sun over a cross-sectional area of πR^2, where R is the average radius of the earth, so the total energy received by the earth is $\pi R^2 S_0$. This energy is shared over the surface of the earth, which has an area of $4\pi R^2$. Thus, the average rate at which solar energy is incident on each unit area of the earth is

$$S_{\text{INC}} = \frac{S_0 \pi R^2}{4\pi R^2} = \frac{S_0}{4} = 342 \text{ W/m}^2. \tag{4.8}$$

Finally, to calculate S_{ABS} we need to take into account that the earth does not absorb all the solar energy incident on it. About 31% of the incident solar radiation is reflected back to space from clouds, molecules, and particulates in the atmosphere, and the surface. The fraction of incident solar radiation reflected by the surface and the atmosphere together is the *planetary albedo*, α. Taking the planetary albedo into account, we have

$$S_{\text{ABS}} = S_{\text{INC}}(1 - \alpha) = \frac{S_0(1 - \alpha)}{4} = 236 \text{ W/m}^2. \tag{4.9}$$

Substituting Eq. 4.9 into Eq. 4.6 allows us to calculate the radiative equilibrium temperature of the earth as

$$T_E = \left[\frac{S_0(1 - \alpha)}{4\sigma} \right]^{1/4} = 254 \text{ K}. \tag{4.10}$$

4.3 HOW CONSTANT IS THE SOLAR CONSTANT?

Although S_0 is called the solar constant, its value changes. Figure 4.2 shows a composite of daily (gray lines) and annual mean (black line) values of the solar constant observed over a period of 35 years by various satellites. The 11-year sunspot cycle is evident, with high values of the solar constant associated with high sunspot counts (solar maximum). Differences in annual mean values of S_0 between solar minimum and solar maximum are about 1 W/m². Day-to-day variations in the solar constant are larger during solar maxima and subdued during solar minima. During solar maxima, the sun's rotation can result in variations in S_0 of up to 4 W/m² (about 0.3%) within the 30-day rotation period as the rotation brings magnetically active regions (sunspots) to face the earth.

In addition to these variations in solar luminosity on daily, monthly, interannual, and decadal time scales associated with solar activity, the solar constant changes on millennial time scales and longer. Variations over tens of thousands of years are associated with the Milankovitch cycles (section 3.5), and solar evolution changes the solar constant over billions of years. According to models of stellar evolution, the sun was significantly cooler billions of years

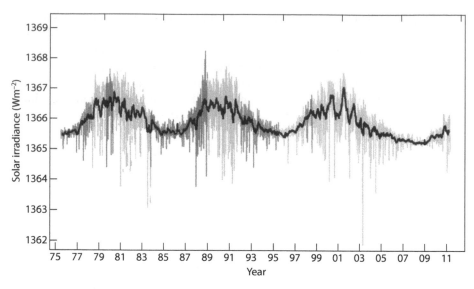

Figure 4.2 Daily (gray) and annually averaged values of the solar constant composited from radiometers on various satellites. See Fröhlich (2006) for the composite methodology. Courtesy of the SOHO consortium, a project of the ESA and NASA.

ago, with a luminosity of about 70% of today's value. According to Eq. 4.10, the earth's radiative equilibrium temperature varies with the one-fourth power of solar luminosity, so the surface temperature of the early earth would have been about $\sqrt[4]{0.70} = 91\%$ of today's value, or about 23 K cooler, assuming no other changes in the system. Climate theory predicts global-scale glaciation for such a decrease in temperature, if all other factors are the same, but there is ample evidence of widespread liquid water (e.g., sedimentary rocks) on the early earth and no evidence of extensive glaciation. This apparent contradiction is known as the *faint young sun paradox*.

Several theories have been advanced to resolve the faint young sun paradox. One is that there were high levels of greenhouse gases in the atmosphere (section 4.5). Another is that less extensive continental surfaces on the earth 2 billion years ago were associated with a lower surface albedo and reduced cloud cover, leading to a significantly lower planetary albedo. Another idea stems from the possibility that Venus was not in orbit about the sun 2 billion years ago, leading to a shorter distance between the earth and the sun and an increase in the solar constant.

4.4 SOLAR AND TERRESTRIAL SPECTRA

The Planck function (Eq. 4.3) with $T = 5500$ K, an emission temperature representative of the sun's photosphere, is drawn in Figure 4.3a. This Planck curve, also known as a *blackbody spectrum*, shows how the emitted energy is distributed across wavelengths for a perfect blackbody with temperature

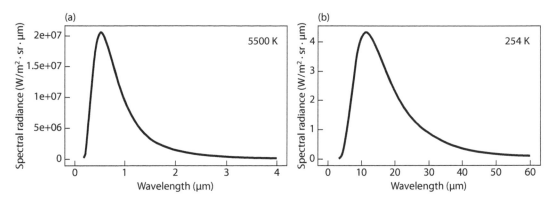

Figure 4.3 Planck curves for temperatures of (a) 5500 K and (b) 254 K.

similar to that of the sun. Consistent with Wein's displacement law (Eq. 4.5), $\lambda_{MAX} \cong 0.5$ mm.

Radiation with wavelengths from about 0.2 to 0.9 μm, ranging between ultraviolet and near-infrared wavelengths, is referred to as *shortwave radiation*. The human eye is "tuned" by evolution to operate near the center of the range of solar radiation, in the *visible spectrum* from 0.4 to 0.7 μm.

The Planck curve for 254 K, the radiative equilibrium temperature of the earth, is plotted in Figure 4.3b. Note that the values of the spectral radiance are about seven orders of magnitude smaller than those for the solar radiation, and wavelengths are longer. At these temperatures, most of the emission is in *infrared* wavelengths, between about 5 and 20 μm, with maximum emission close to 11 μm. Emission in this range of wavelengths is termed *longwave* or *terrestrial* radiation. It is also referred to as *thermal emission* because it is sensed by skin as heat.

The Planck curves of Figure 4.3 are *emission spectra* for the earth and the sun predicted using blackbody theory. How closely do the observed emission spectra match these idealized spectra?

The solar *spectral irradiance* observed from outside the earth's atmosphere is shown in Figure 4.4 by the lighter gray shading. Spectral irradiance is the energy *received* by a sensor, in contrast to the radiance which is the energy *emitted* according to Planck's law (Eq. 4.3). The black line is the irradiance for a blackbody at 5500 K. It is apparent that blackbody theory provides a good approximation to the observed emission spectrum. Just under half the sun's total emitted energy is in visible and near-infrared wavelengths. The solar emission in ultraviolet wavelengths is smaller than predicted by blackbody theory and greater in the X-ray, far-ultraviolet, and radio wavelengths. These deviations from an ideal blackbody spectrum occur primarily because of temperature variations across the *photosphere* (the level from which radiation is emitted from the sun) and because of absorption and emission by constituents within the solar atmosphere.

The other spectral irradiance curve displayed in Figure 4.4, indicated by the darker gray shading, is observed from the earth's surface, after the incoming solar radiation has passed through the atmosphere. This spectrum deviates

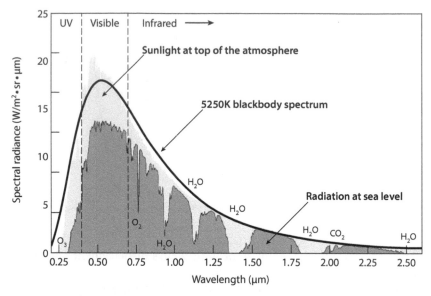

Figure 4.4 Solar spectral irradiance observed by a satellite outside the earth's atmosphere (light gray) and observed at the surface (dark gray).

significantly from the solar spectrum observed outside the atmosphere. The shape is similar to that of an ideal blackbody, but it is clear that when solar radiation passes through the earth's atmosphere, energy is removed from the solar spectrum across all wavelengths, and the loss is particularly strong and even complete at certain wavelengths.

The energy lost from the solar spectrum across a broad range of wavelengths is due to reflection and scattering, which change the direction of the beam of energy but not its wavelength. Two scattering processes are important in the earth's atmosphere:

(1) *Rayleigh scattering* by atmospheric molecules and small particles (with diameters $< 0.1\ \mu m$) is the dominant form of scattering. This process is most effective at shorter visible wavelengths, that is, the blue region of the visible spectrum, since the amount of energy scattered is proportional to λ^{-4}. Rayleigh scattering removes blue visible light from the solar beam and scatters it across the sky, which makes the sky appear blue. Most of the energy that Rayleigh scatters from a molecule is redirected either forward or backward.

(2) *Mie* (pronounced "me") *scattering* occurs when solar radiation interacts with larger particles in the atmosphere, with diameters on the order of (or larger than) the wavelength of the radiation. Examples are dust, pollen, smoke, and water droplets. Mie scattering does not have a strong wavelength dependence so the scattered light is white. Because of Mie scattering, clouds, mist, and fog appear white, and there is a white glare around the sun when numerous water droplets are suspended in the air. Mie scattering is predominantly forward scattering and becomes more so as the particle size increases.

Loss of energy from the solar spectrum at particular wavelengths occurs because of *molecular absorption*. Because of their structure, molecules absorb radiation at particular wavelengths, and for some molecules in the atmosphere these wavelengths are within the solar spectrum. For example, an ozone molecule (O_3) has three oxygen atoms arranged in an isosceles triangle in the two-dimensional plane. Only two sides of the ozone triangle are occupied by chemical bonds, so two of the atoms are not bound to each other. With this structure, the ozone molecule has three vibrational degrees of freedom (symmetric stretching, bending, and asymmetric stretching) and three rotational degrees of freedom (one each around the x-, y-, and z-axes). As a result, ozone absorbs radiation at certain wavelengths—actually, in certain wavelength bands, since temperature and precipitation variations broaden the absorption bands—that excite these allowed modes of motion, removing the energy from the radiation spectrum at those certain wavelengths and converting it into the kinetic energy of vibration or rotation.

Ozone absorption is strong in the ultraviolet part of the spectrum. Note in Figure 4.4 that the solar radiation incident at the top of the atmosphere is depleted in wavelengths shorter than about 0.3 μm by the time it reaches the ground. There is also significant absorption at visible wavelengths by ozone, in the Chappuis bands from 0.44 to 0.74 μm. Atmospheric water vapor (H_2O) is responsible for the broad absorption bands centered near 0.9 μm, 1.1 μm, 1.4 μm, and 1.9 μm. Molecular oxygen (O_2) and carbon dioxide (CO_2) also contribute to shortwave absorption. Table 4.1 lists many of the principal absorbing molecules in the earth's atmosphere and the wavelengths at which they absorb radiation.

Observed spectra of longwave emission from the earth are shown in Figure 4.5. The black curve is the observed emission at the surface, before the energy passes through the atmosphere. The jagged gray line is earth's spectral irradiance recorded by a satellite outside the atmosphere. This radiation is known as the *OLR*, or *outgoing longwave radiation*.

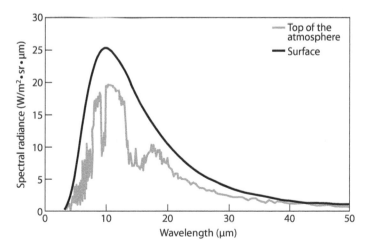

Figure 4.5 Terrestrial emission of radiation measured at the surface (black line) and at the top of the atmosphere (gray line). Adapted from Kiehl and Trenberth (1997).

Table 4.1. Absorption of solar and terrestrial radiation by atmospheric molecules

Absorbing molecule	Spectral regions (μm)[a]		
	Solar wavelengths ~0.1–2.5 μm	Terrestrial wavelengths ~2.5–25 μm	Approximate atmospheric residence time[b]
Carbon dioxide (CO_2)	1.4, 1.6, 2.0, 0.78–1.24 (weak)	13.5–16.5, centered at 15 4.2–4.4, centered at 4.3 2.7, 5.2, 9.4, 10.4	5–200 years
Water vapor (H_2O)	0.72, 0.81, 0.94, 1.1, 1.4, 1.9	5.5–7.5, centered at 6.3 2.6–3.3 Water vapor continuum[c]	10 days
Ozone (O_3)	0.18–0.34, centered at 0.26 0.32–0.36, 0.44–0.74	2.7, 3.27, 3.59, 4.75, 5.75, 9.0 9.6, 14.1	2 months
Methane (CH_4)		3.3, 7.7	10 years
Nitrous oxide (N_2O)		4.5, 7.8, 17.0	100 years
Carbon monoxide (CO)	1.19, 1.57, 2.35	4.67 2.38–25.0	A few months

[a] Only absorption at wavelengths relevant for the climate system are listed. In general, these molecules will also have absorption lines and bands in other wavelengths, but this absorption does not directly influence the flow of radiative energy through the atmosphere.

[b] Residence times vary according to the atmospheric conditions. These values are estimates.

[c] Water vapor absorption is complicated (see text) because the combined vibrational and rotational modes give rise to tens of thousands of absorption lines.

As with the solar spectrum, energy is absorbed from the terrestrial spectrum (see Table 4.1). Molecules in the earth's atmosphere that absorb longwave radiation are called *greenhouse gases*.[2] The most abundant greenhouse gas in the earth's atmosphere is water vapor, and it is responsible for about 70% of the atmospheric absorption of terrestrial radiation globally. Combined vibrational and rotational modes give rise to tens of thousands of water vapor absorption lines. Gas, liquid, and solid states of water are all found in the atmosphere and have different absorption characteristics. *Continuum absorption* occurs throughout the infrared and microwave portions of the spectra. This

[2] Some greenhouse gases also absorb shortwave radiation. Ozone, for example, is radiatively active in both solar and terrestrial wavelengths

absorption is not highly dependent on wavelength and occurs in addition to line absorption.

Carbon dioxide is also a key greenhouse gas, responsible for about 25% of the global atmospheric absorption of terrestrial radiation. CO_2 absorption between 13.5 and 16.5 μm is particularly relevant, since these wavelengths are near the peak in the terrestrial emission spectrum (see Figs.4.3b and 4.5). The pronounced divot near 9.6 mm in the spectrum in Figure 4.5 is due to O_3 absorption. Also apparent in the earth's emission spectrum are absorption lines of methane and nitrous oxide.

The atmosphere is relatively transparent to radiation at certain wavelengths, in regions of the spectrum called *atmospheric windows*. A prominent longwave window is found between about 8 μm and 13 μm, bordered by H_2O and CO_2 absorption bands and interrupted by the strong O_3 absorption line at 9.6 μm. Most of the terrestrial radiation that emerges from the top of the atmosphere has wavelengths in this range. Instruments mounted on satellites take advantage of these windows to observe the earth system. The longwave IR window channel, for example, is at 10.7 μm. Energy emitted from the surface at this wavelength passes through the atmosphere with minimal absorption by molecules in the atmosphere, so the observed emission temperature represents the actual surface temperature.

Emission spectra can be used to infer vertical distributions of atmospheric constituents when the temperature profile is known, or to infer the temperature profile when the vertical distribution of a constituent is known. For example, Figure 4.6 shows a terrestrial spectrum observed over the tropical Pacific by an infrared detector on the Nimbus 4 satellite. Here, both wavelength and wavenumber are indicated on the x-axis. (Wavenumber, k, is $2\pi/\lambda$.) The dashed lines are Planck curves for various temperatures. By matching a segment of the observed spectrum with a Planck curve, one can assign an emission temperature to that part of the spectrum. For example, the energy emitted in the wavelength range 11–13 μm is being emitted at about 295 K. The fact that this is the atmospheric longwave window, and the temperature is high, suggests that this emission originates from the surface.

Examine the longwave spectrum in the vicinity of the 15 μm CO_2 absorption band. This section of the spectrum fits the Planck curve for a temperature of about 215 K. According to Figure 2.8, a temperature of 215 K is characteristic of an elevation of about 10 km, so the CO_2 emissions detected by the satellite originate in the upper tropopause. Emission by H_2O occurs at about 270 K, in the lower troposphere, since the sections of the spectrum at which H_2O is radiatively active (e.g., near 20 μm and 7.5 μm) approximate the blackbody curve for that temperature. This is consistent with Figure 2.30, which shows that water vapor in the atmosphere is primarily located in the lower troposphere. In contrast, CO_2 is well mixed in the atmosphere.

Another way to represent the influence of various constituents on the flow of longwave and shortwave radiation through the atmosphere is by constructing *absorption spectra*. In contrast to the *emission spectra* shown in Figures 4.3–4.6, an absorption spectrum displays the fraction of the radiation at a given wavelength that is absorbed as a function of wavelength—that is, the absorptivity—as that radiation passes through the atmosphere. Figure 4.7a

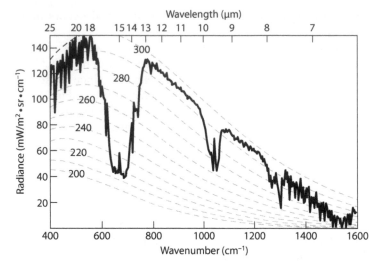

Figure 4.6 Emission spectrum of the earth observed at the top of the atmosphere, with Planck curves for various temperatures drawn. Adapted from Hanel et al. (1972).

shows Planck curves (emission spectra) for temperatures representative of solar and terrestrial temperatures. Note the scaling of the axes needed to display these two spectra on one plot. It is important to remember that molecular absorption is irrelevant to the climate system if it occurs at wavelengths not covered by these spectra.

Figure 4.7b shows absorption spectra for the stratosphere (top) and the entire atmosphere (bottom). The atmosphere is more transparent at solar wavelengths than at terrestrial wavelengths. For wavelengths shorter than about $0.3 \, \mu m$ and longer than about $22 \, \mu m$, the atmosphere is opaque.

The panels of Figure 4.7c portray absorption spectra for individual molecules. The opacity of the atmosphere at long wavelengths ($\lambda > 22 \, \mu m$) is due to absorption by H_2O, and its opacity at short wavelengths ($\lambda < 0.3 \, \mu m$) is due to O_3, as discussed above. Strong absorption near the peak of the terrestrial spectrum is due to CO_2. Methane absorbs approximately 90% of the energy at $3.3 \, \mu m$ and $7.7 \, \mu m$. N_2O absorption removes more than 90% of the radiation near $4.5 \, mm$ and $7.8 \, \mu m$, and the O_3 absorption band at $9.6 \, \mu m$ is also powerful.

4.5 THE GREENHOUSE EFFECT

Absorption of longwave radiation by greenhouse gases in the atmosphere redistributes heat within the climate system, warming the surface and lower troposphere by a process known as the *greenhouse effect*.

The *slab atmosphere model* is an idealized version of the earth/atmosphere system that can be used to demonstrate the greenhouse effect, and allows us to make quantitative estimates of the influence of the greenhouse effect on atmospheric and surface temperatures. The spherical geometry of the earth is removed from the problem, and the surface is modeled as a semi-infinite plane

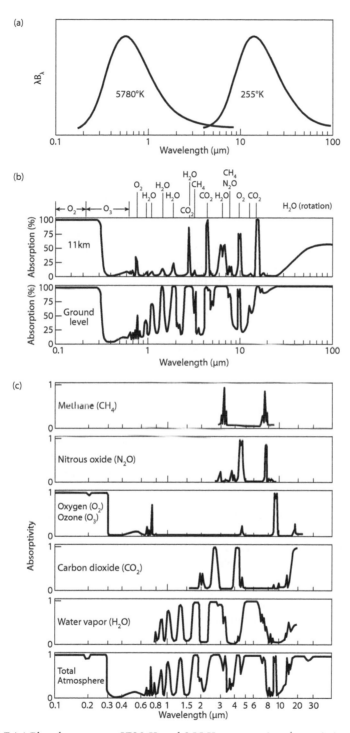

Figure 4.7 (a) Planck curves at 5780 K and 255 K representing the emission spectra of the sun and the earth, respectively. The y-axis is scaled by multiplying radiance (B_λ) by wavelength so the two spectra can be plotted on the same figure (see Exercise 4.2). (b) Absorption spectra for the stratosphere (top) and the entire atmosphere (bottom). (c) Absorption spectra for the most important absorbers in the earth's atmosphere.

overlain by a semi-infinite plane atmosphere. Then, Eq. 4.9 is used to calculate the solar energy absorbed by the slab atmosphere system as 236 W/m².

Several cases of slab atmosphere calculations are presented below to illustrate various aspects of the greenhouse effect.

GREENHOUSE CASE I: NO ATMOSPHERE

The simplest case is one with no atmosphere. In this case, the incoming solar radiation is absorbed at the surface, as illustrated in Figure 4.8.

Assume that the surface behaves as a perfect blackbody, absorbing and reemitting all this shortwave radiation. The surface is represented as being insulated on its lower side to account for the fact that radiation does not penetrate far into the earth's surface, and the longwave radiation is emitted back to the atmosphere (rather than down into the earth). Assuming radiative equilibrium, so that radiative heating and radiative cooling are balanced and the surface temperature, T_S, does not change gives

$$S_{ABS} = \sigma T_S^4 \Rightarrow T_S = 254 \text{ K}. \tag{4.11}$$

This is the same calculation as in Eq. 4.10. In the "no atmosphere" case, the surface temperature is T_E.

GREENHOUSE CASE II

Add a slab atmosphere over the slab surface, shown in Figure 4.9. Even though the slab atmosphere touches the surface, it is drawn suspended over the surface so the fluxes between the atmosphere and the surface can be clearly depicted. Assume the following:

- The surface and atmosphere are both perfect black bodies in radiative equilibrium.
- The atmosphere is transparent to solar radiation.
- The atmosphere, like the surface, is opaque to longwave radiation and absorbs all longwave (terrestrial) radiation incident on it.

The slab atmosphere, unlike the surface, emits radiation both upward to space and downward back to the surface.

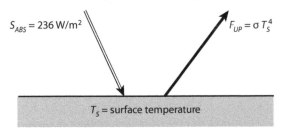

$S_{ABS} = 236 \text{ W/m}^2$ $F_{UP} = \sigma T_S^4$

T_S = surface temperature

Figure 4.8 Greenhouse slab model, Case I.

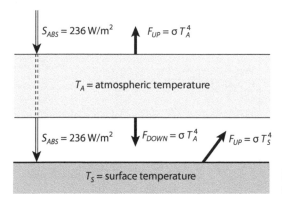

Figure 4.9 Greenhouse slab model, Case II.

Use the radiative equilibrium condition to generate a set of two equations and two unknowns, namely, T_S and T_A, by setting the heat input equal to the heat output for the surface slab and for the atmospheric slab:

For the atmosphere,

$$\sigma T_S^4 = 2\sigma T_A^4. \tag{4.12}$$

For the surface,

$$S_{ABS} + \sigma T_A^4 = \sigma T_S^4. \tag{4.13}$$

Solving Eqs. 4.12 and 4.13 simultaneously gives $T_S = 302$ K, and $T_A = 254$ K. Note that the OLR from the system is coming from the atmosphere, not the surface, so the radiative balance for the entire system is

$$S_{ABS} = \sigma T_A^4 \Rightarrow T_A = 254 \text{ K}. \tag{4.14}$$

The surface emits longwave radiation at a much higher temperature than that of the atmosphere, which is at the radiative equilibrium temperature.

This simple model illustrates the essence of the greenhouse effect. The atmosphere absorbs longwave radiation emitted by the surface and reradiates that energy at the atmospheric temperature. This reradiation directs additional energy back to the surface. Thus, the surface has two heat sources—the solar radiation and the *longwave back radiation* from the atmosphere. The surface is much warmer as a result. Meanwhile, the OLR from the earth system originates in the atmosphere, so the atmospheric temperature is maintained at the value that balances the solar radiation absorbed by the system, that is, T_E.

GREENHOUSE CASE III

A slightly more sophisticated, and realistic, assumption is that the atmosphere absorbs only a fraction, f, of the longwave radiation emitted from the surface.

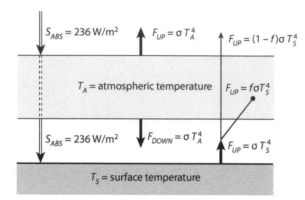

Figure 4.10 Greenhouse slab model, Case III.

As indicated in Figure 4.10, the heat balances for the atmosphere and the surface, respectively, yield the following two equations:

$$f\sigma T_S^4 = 2\sigma T_A^4 \tag{4.15}$$

$$S_{ABS} + \sigma T_A^4 = \sigma T_S^4. \tag{4.16}$$

Then,

$$\sigma T_S^4 = \left(\frac{2}{2-f}\right) S_{ABS}. \tag{4.17}$$

Note that this case reduces to Case I if $f = 0$ and to Case II if $f = 1$. When more longwave radiation is absorbed by the atmosphere, that is, when the greenhouse effect is more powerful, f is larger and the surface temperature is higher.

GREENHOUSE CASE IV

Assume that rather than being transparent to solar radiation, the atmosphere absorbs 10% of the solar radiation and 100% of the longwave radiation. In that case, which is drawn in Figure 4.11, the heat balance equations for the atmosphere and the surface are,

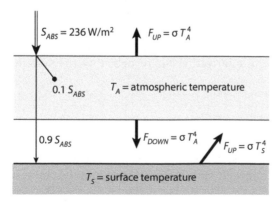

Figure 4.11 Greenhouse slab model, Case IV.

$$0.1S_{ABS} + \sigma T_S^4 = 2\sigma T_A^4 \tag{4.18}$$

and

$$0.9S_{ABS} + \sigma T_A^4 = \sigma T_S^4. \tag{4.19}$$

Solving Eqs. 4.18 and 4.19 simultaneously, we get $T_S = 298$ K and $T_A = 254$ K. Again, T_A is the same as in Case II, in which no solar radiation is absorbed by the atmosphere. The radiative balance at the top of the atmosphere must be preserved—the OLR ($F_{UP} = \sigma T_A^4$ in the slab atmosphere model) must always balance S_{ABS}, so T_A cannot change unless S_{ABS}, changes. But since the atmosphere is now absorbing some solar radiation, it must absorb less terrestrial radiation from the surface to keep its temperature the same as in Case II. Since the atmosphere is opaque to the terrestrial radiation, the only way for the atmosphere to absorb less longwave radiation is for the surface temperature to be lower than in Case II.

GREENHOUSE CASE V

A major shortcoming of the preceding examples is that the atmosphere is assumed to be isothermal, since the single slab has a uniform temperature. We know, however, that temperature falls off at a rate of approximately 6 K/km in the troposphere (Fig. 2.8). Changes in atmospheric temperature with height can be taken into account in the simple slab model formulation by using more than one atmospheric slab. Consider, for example, the two-slab case shown in Figure 4.12, in which each atmospheric layer is assumed to be transparent to solar radiation and opaque to longwave radiation. In this case, the analytical solution requires solving a system of three equations and three unknowns:

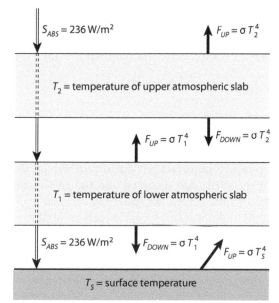

Figure 4.12 Greenhouse slab model, Case V.

For the surface,

$$S_{ABS} + \sigma T_1^4 = \sigma T_S^4. \tag{4.20}$$

For atmosphere layer 1,

$$\sigma T_S^4 + \sigma T_2^4 = 2\sigma T_1^4. \tag{4.21}$$

For atmosphere layer 2,

$$\sigma T_1^4 = 2\sigma T_2^4. \tag{4.22}$$

Simultaneous solution of Eqs. 4.20–4.22 results in $T_S = 334$ K (very warm!), $T_1 = 302$ K (same as the surface temperature in the single-slab Case II), and $T_2 = 254$ K. The longwave emission from the top of the atmosphere is at the radiative equilibrium temperature ($T_2 = T_E$), the emission temperature needed to balance the energy absorbed by the system.

The simple slab atmosphere calculations illustrate the redistribution of heat that occurs in the climate system due to the greenhouse effect while maintaining the state of radiative equilibrium between the earth and the sun. In the actual atmosphere, increases in greenhouse gases lead to tropospheric and surface warming, but the stratosphere cools when greenhouse gas levels increase. The effects are different in the troposphere and stratosphere because the radiative sources and sinks of heat are different. In the troposphere, the atmosphere is warmed primarily by greenhouse gas absorption of longwave radiation and cooled by the emission of longwave radiation by those same greenhouse gases. In contrast, the major source of heating in the stratosphere is the absorption of shortwave (ultraviolet) radiation by ozone. (As discussed in chapter 2, this is the reason for the local temperature maximum at the stratopause seen in Fig.2.8.) This heat is transferred to the other molecules in the stratosphere, including greenhouse gas molecules, in collisions. A heated greenhouse gas molecule will then radiate longwave radiation and cool the stratosphere. Thus, when there are more greenhouse gas molecules in the stratosphere, there is more cooling. An added effect of increased greenhouse gases in the troposphere is that they reduce the upward longwave flux from the troposphere into the stratosphere (see chapter 10).

4.6 THE EQUATION OF TRANSFER

The *equation of transfer* is used to calculate changes in the intensity of radiation as it travels through the atmosphere. Because transit distances through the atmosphere are much smaller than the earth–sun distance and the earth's radius, the $1/r^2$ dependence of radiation emitted from the spherical sun and from the earth's surface is negligible and the radiation can be accurately treated as if the beams of radiation were parallel.

The spectral radiance (defined in section 4.1) can increase or decrease as a beam of radiation passes through the atmosphere. The radiance at a given wavelength decreases when molecules and particles in the atmosphere absorb

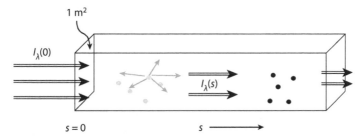

Figure 4.13 Attenuation of a beam of radiation passing through a medium with scattering particles (gray) and absorbing molecules (black).

energy and reradiate at a different wavelength, and when molecules and particles scatter energy from the path of the beam. Spectral radiance can increase due to emission from molecules and particles along the path, and when scattering directs energy into the path.

Consider the one-dimensional problem shown in Figure 4.13, in which the spectral radiance, $I_\lambda(s)$, is changed by absorption, emission, and scattering along a path length s. First, consider the attenuation of the beam due to the absorption of energy by molecules acting as blackbodies. The *absorption coefficient*, k, is defined as the fraction of incident radiant energy absorbed per unit path length, or

$$k = -\frac{dI_\lambda/I_\lambda}{ds} = -\frac{1}{I_\lambda}\frac{dI_\lambda}{ds}. \tag{4.23}$$

Rearranging Eq. 4.23, we have the attenuation of the beam due to absorption:

$$\left(\frac{dI_\lambda}{ds}\right)_{\text{ABS}} = -kI_\lambda. \tag{4.24}$$

Now, consider increases in the spectral radiance of the beam due to thermal emission, which is governed by the Planck function, B_λ (Eq. 4.3). According to Kirchhoff's law (section 4.1), the emissivity is equal to the absorptivity, so

$$\left(\frac{dI_\lambda}{ds}\right)_{\text{EMIS}} = +kB_\lambda. \tag{4.25}$$

The total change in spectral radiance for a beam traveling through a non-scattering medium is then

$$\frac{dI_\lambda}{ds} = \left(\frac{dI_\lambda}{ds}\right)_{\text{ABS}} + \left(\frac{dI_\lambda}{ds}\right)_{\text{EMIS}} = k(B_\lambda - I_\lambda). \tag{4.26}$$

Equation 4.26 is known as *Schwarzschild's equation*. It states that a beam of light will be attenuated in passing through a medium (with no scattering present) if $I_\lambda > B_\lambda$ and it will be enhanced if $I_\lambda < B_\lambda$.

If $I_\lambda >> B_\lambda$, so emission into the path of the beam is negligible, the solution to Eq. 4.26 is

$$I_\lambda = I(0)e^{-ks}, \tag{4.27}$$

where $I_\lambda(0)$ is the intensity of the beam at $s = 0$ (see Fig. 4.13). According to Eq. 4.27, the intensity of the beam decreases exponentially with path length due to molecular absorption. If there are N_{ABS} absorbers per unit volume, and each has a cross section denoted by A_{ABS} and an absorptivity of 1, then

$$k = N_{ABS} A_{ABS}. \tag{4.28}$$

Note that k has units of m^{-1}.

Optical depth, τ, is a dimensionless quantity that measures the opacity of the atmosphere (or any medium), or the penetration depth of a beam of radiation into the medium. It is defined as

$$\tau \equiv ks, \tag{4.29}$$

so that

$$I_\lambda = I_\lambda(0)e^{-\tau} \tag{4.30}$$

for the case of beam attenuation by molecular absorption. If $\tau = 0$, the medium is completely transparent to the radiation, and $I_\lambda(s) = I_\lambda(0)$. When $\tau = 1$, the intensity of the beam is reduced to about one-third $(1/e)$ of its original intensity, since, from Eq. 4.30,

$$I_\lambda = \frac{I_\lambda(0)}{e} \approx 0.3 I_\lambda(0) \tag{4.31}$$

when $\tau = 1$.

A similar approach allows us to express the loss of energy from a beam due to scattering. If there are N_{SCAT} scatterers, each with area A_{SCAT}, then the attenuation of the beam by scattering is

$$\frac{dI_\lambda}{ds} = -f_{SCAT} N_{SCAT} A_{SCAT} I_\lambda, \tag{4.32}$$

where the factor f_{SCAT} is the fraction of the scattering that is not forward scattering. For example, f_{SCAT} will be larger for the Rayleigh scattering of shorter wavelengths than for Mie scattering.

Assuming that the density of absorbers, emitters, and scatterers in the volume is sufficiently low so there is no significant interference among these processes, the various influences on the beam intensity can be linearly superimposed to form the equation of transfer

$$\frac{dI_\lambda}{ds} = k(B_\lambda - I_\lambda) - f_{SCAT} N_{SCAT} A_{SCAT} I_\lambda. \tag{4.33}$$

4.7 RADIATIVE EFFECTS OF CLOUDS

Clouds interact with both shortwave and longwave radiation in the earth's atmosphere. Their most important interaction with shortwave radiation is by reflection, which changes the direction of the incident radiation but not its wavelength. Cloud albedos range from 0.2 to 0.9 (see Table 5.1). Some of this reflected radiation is redirected back into space, making the climate system cooler. At the same time, clouds behave as nearly perfect blackbodies at terrestrial wavelengths. They absorb almost all the longwave radiation incident on them, and reradiate the energy at different wavelengths according to their temperatures and the Stefan-Boltzmann law (Eq. 4.1). If the longwave radiation incident on a cloud was originally directed out of the atmosphere, for example, as upward emission from the surface, then the cloud will reemit this radiation in all directions and at lower temperatures, turing the energy back into the climate system to warm it. This, of course, is just the greenhouse effect at work.

Which of these cloud effects on radiation dominates for the earth system? Do clouds cool the earth system, or warm it? The two idealized cases that follow make it clear that the answer to this question depends on factors such as the cloud's albedo and its altitude.

CLOUD CASE 1

Consider a low cloud with a shortwave albedo of 0.3 and a longwave emissivity (and absorptivity) of 1. To isolate the effects of this one cloud, assume that the rest of the atmosphere above and below the cloud is empty of other clouds, greenhouse gases, and particles, and model the cloud as a semi-infinite slab. The cloud top is located near 2 km elevation, so a reasonable value for the cloud-top temperature, T_C, is

$$T_C = T_S + 2\Gamma = 287.5 \text{ K} - (2 \text{ km})(6 \text{ K/km}) = 275.5 \text{ K}, \qquad (4.34)$$

where the globally averaged surface air temperature and a typical tropospheric lapse rate are used. Further assume that the solar radiation incident on the cloud, S_{INC}, is also the globally averaged value of 342 W/m^2 (Eq. 4.8).

What is the net radiative effect of this cloud?

- Shortwave effect: The cloud reflects 30% of the incoming solar radiation. Because of this reflection—because of the presence of the cloud—the climate system receives 102.6 W/m^2 less radiative heating.
- Longwave effect: The cloud absorbs all the $\sigma T_S^4 = 387.4$ W/m^2 longwave radiation incident on it from the surface and emits thermal radiation of $\sigma T_C^4 = 326.6$ W/m^2. Because of the presence of the cloud, the climate system retains 60.8 W/m^2 more heat

Thus, the net effect of the presence of this low cloud is to cool the climate system by 41.8 W/m^2.

CLOUD CASE 2

Consider a high cloud with the exact same radiative characteristics as the low cloud in Case 1, that is, a shortwave albedo of 0.3 and longwave emissivity of 1. Again, assume that the rest of the atmosphere above and below the cloud is empty of other clouds, greenhouse gases, and particles. The cloud top is located near 10 km elevation, so a reasonable value for its temperature is

$$T_C = T_s + 2\Gamma = 287.5 \text{ K} - (10 \text{ km})(6 \text{ K/km}) = 227.5 \text{ K}. \qquad (4.35)$$

What is the net effect of this high cloud on the heat balance?

- Shortwave effect: The cloud interacts with the solar radiation in the same way as the low cloud because the albedo is independent of the cloud height and temperature, and the climate system receives 102.6 W/m^2 less solar heating.
- Longwave effect: The cloud again absorbs the entire $\sigma T_S^4 = 387.4$ W/m^2 longwave radiation incident on it from the surface and emits thermal radiation of $\sigma T_C^4 = 151.9$ W/m^2. The decrease in longwave radiation emitted to space due to the presence of the high cloud is 235.5 W/m^2.

In the high-cloud case, the reduction in OLR due to the presence of the cloud is greater than the reduction in solar heating by 132.9 W/m^2. The net effect of the presence of the high cloud is to *warm* the climate system by 132.9 W/m^2.

These simple calculations demonstrate that, to first order, interactions between clouds and solar radiation are independent of cloud height, but interactions between clouds and longwave radiation depend on cloud temperature and, therefore, cloud altitude. High clouds are much more effective at trapping longwave radiation in the climate system than low clouds because the temperature difference between high clouds and the surface is greater. For this reason, high clouds tend to warm the climate system and low clouds tend to cool the system.

For the earth system as a whole, we assumed previously that there is a balance between the incoming solar radiation and the outgoing longwave radiation. Locally, however, this state of radiative equilibrium does not hold in general. For the cloudless case, the net radiative heating at a particular location is given by

$$H_{CS} = \frac{(1 - \alpha_{CS})S_0}{4} - \sigma T_{CS}^4, \qquad (4.36)$$

where H_{CS} is the "clear-sky" heating rate, α_{CS} is the clear-sky albedo, and T_{CS} is the atmospheric temperature. Similarly, the net radiative heating in a cloudy atmosphere, H_{CL}, is

$$H_{CL} = \frac{(1 - \alpha_{CL})S_0}{4} - \sigma T_{CL}^4. \qquad (4.37)$$

Cloud forcing, F_C, is the net local radiative heating rate due to the presence of clouds, so it is the difference in radiative heating between the cloudy and clear-sky cases:

$$F_C \equiv H_{CL} - H_{CS} = \frac{(\alpha_{CS} - \alpha_{CL})S_0}{4} + \sigma(T_{CS}^4 - T_{CL}^4). \qquad (4.38)$$

Because $\alpha_{CS} < \alpha_{CL}$, the first term on the right-hand side of Eq. 4.38 is negative, indicating that shortwave cloud forcing is always negative, a cooling effect. The second term on the right-hand side is the longwave cloud forcing. In the absence of clouds, most of the thermal emission from the climate system comes from atmospheric water vapor and, as seen in Fig. 2.30, nearly all the water vapor is located in the lower troposphere—below about 800 hPa, or about 2 km. Therefore, $T_{CS}^4 > T_{CL}^4$, and the longwave cloud forcing is positive, a heating effect. The sign of F_C depends on which of these two terms is larger.

Estimates based on satellite observations indicate that, in the global average, the magnitude of the longwave forcing is 31.1 W/m^2, and the magnitude of the shortwave forcing is 48.4 W/m^2. From these measurements, then, the presence of clouds cools the climate system ($F_C = -17.3$ W/m^2). As discussed in chapter 11, this does not inform us about the role of clouds as climate changes.

Locally, clouds can have either a heating or a cooling effect. Positive values of F_C are observed over the western Pacific warm pool, for example, where deep convective clouds form. Negative values of F_C are observed in association with the marine stratus clouds that form at low levels over the cold-tongue regions of the Pacific and Atlantic. Cloud forcing in these regions helps maintain the cold surface waters of the eastern ocean basins.

4.8 REFERENCES

Fröhlich, C., 2006: Solar irradiance variability since 1978. Revisions of the PMOD composite during solar cycle 21. Space Science Reviews, *125*, 53–65.

Goody, R. A., and G. D. Robinson, 1951: "Radiation in the troposphere and lower stratosphere." *Quarterly Journal of the Royal Meteorological Society* 77: 153.

Hanel, R. A., et al., 1972: "The Nimbus 4 infrared spectroscopy experiment." *Journal of Geophysical Research* 77: 2629–2641.

Kiehl, J. T., and K. T. Trenberth, 1997: "Earth's global mean energy budget." *Bulletin of the American Meteorological Society* 78: 197–208.

4.9 EXERCISES

4.1. The temperature of the blackbody curve that most closely matches an observed spectrum is called the *brightness temperature* of the emitting surface. Using Wein's displacement law, estimate the brightness temperature for the sun and for the earth from an examination of their spectra in Figures 4.4 and 4.5.

4.2. Sketch the emission spectrum from the sun at the top of the atmosphere, using the same y-axis as in Figure 4.4 but extending the x-axis to 40 μm. Now, try to draw the terrestrial spectrum observed at the surface (see Fig. 4.5) to scale on your graph.

4.3. Calculate the solar constant for Mars.

4.4. Consider a unit area (1 m^2) flat plate sitting in space at a distance from the sun equal to the average earth–sun distance and oriented perpendicular to the solar beam. The back of the plate is insulated, so emission from the plate can come only from the side facing the sun. Assuming that the plate behaves as an idealized blackbody with emissivity of 1, what will be the temperature of the plate? What will be the temperature of the plate if the back is not insulated? What will be the temperature if the insulated plate is oriented at a 45° angle to the solar beam?

4.5. The atmosphere of Mars is much thinner than that of Earth, with surface pressure of only 8 hPa on average, but it is about 95% CO_2, a greenhouse gas. About 10% of the longwave radiation from the surface is absorbed by the Martian atmosphere. Use the slab-atmosphere greenhouse model to estimate the radiative equilibrium temperature of Mars and its surface temperature. Assume that the planetary albedo is 0.15.

4.6. Consider solar radiation of wavelength λ, assumed to be in parallel beams, incident at the top of a slab atmosphere with intensity I_0. This idealized atmosphere contains a uniform distribution of molecules that absorb all the radiation at wavelength λ incident on them. These molecules have a radius of 5 μm and their number density is 3×10^6 m^{-3}.

By what percentage will the original intensity of the beam be reduced due to absorption after it travels 2 km through this atmosphere? At what distance from the top of the atmosphere is the optical depth equal to 1? (Neglect scattering and thermal emission.)

4.7. (a) During a warm ENSO event, sea surface temperatures increase by about 3 K in the central equatorial Pacific Ocean. For the clear-sky case, what change (sign and magnitude) in OLR over the central Pacific accompanies this warming?

(b) Also during a warm ENSO event, deep convection moves from the western Pacific to the central Pacific. Assuming that there are no clouds over the central Pacific when a warm event is not in progress, what change in OLR over the central Pacific accompanies this change in the distribution of deep convection? (Assume that the convective clouds cover the full depth of the troposphere.)

(c) What magnitude and sign of OLR anomaly would you expect to observe over the central equatorial Pacific during an ENSO event?

THERMODYNAMICS AND THE FLOW OF HEAT THROUGH THE CLIMATE SYSTEM

Studying the thermodynamics of the atmosphere and oceans provides an understanding of how the solar radiation that fuels the climate system is converted into other forms of energy within the system. The *first law of thermodynamics*, which states that energy can be converted from one form to another but never created or destroyed, provides a foundation from which to study the flow of heat through the climate system and to understand how that heating is related to the circulation.

5.1 EQUATIONS OF STATE

State variables describe a thermodynamic system, and the relationship among them is the *equation of state*. For the atmosphere, the state variables are pressure (p), density (ρ), and temperature (T). The ideal gas law serves as an accurate equation of state. We will use it in the form

$$p = \rho RT, \qquad (5.1)$$

where R is the gas constant for dry air, 287 J/(kg · K). According to Eq. 5.1, the thermodynamic state of a parcel of air is fully determined if two of the three state variables are known.

The equation of state for ocean water is more complicated. There are four state variables, namely, pressure, water density, temperature, and salinity (p, ρ_w, T, and S), and the relationship among them is not simple. A graphical method is often used to determine the density of seawater.

Figure 5.1 is a *T–S plot*, which relates the density of seawater to temperature and salinity. For the observed ranges of sea surface temperatures (Fig. 2.15), about 270 K (–3°C) to 303 K (30°C), and sea surface salinities, about 31–38 psu (Fig. 2.19), seawater density ranges from roughly 1020 kg/m³ to 1030 kg/m³. The letter M indicates the approximate location of Mediterranean waters on the *T–S* plot, and the letter A indicates Arctic waters.

The curvature of the *isopycnals*, or lines of constant density, in the *T–S* plot indicates that the relationship between seawater density and temperature is not linear. Figure 5.2a shows the dependence of the density of fresh water ($S = 0$) on temperature. As water is cooled from 20°C its density increases until

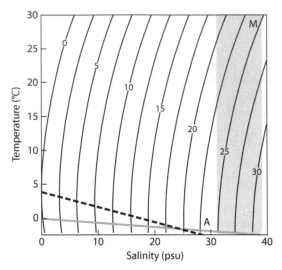

Figure 5.1 *T–S* plot showing the dependence of seawater density on temperature and salinity. Contours of constant density are labeled $\sigma = \rho_w - 1000$, where ρ_w is the density of seawater (kg/m³). Gray shading indicates approximate temperature and salinity ranges of ocean surface waters. "M" indicates the location of Mediterranean waters, and "A" the location of Arctic waters. The dashed line indicates the density maxima, and the gray line denotes the freezing temperature.

it reaches a temperature of 3.98°C. Further cooling of the water *decreases* its density. This unusual behavior of water is the reason that ice forms on the surface of lakes and, more generally, why ice floats. Because lakes freeze from the top, the habitat for fish and other lacustrine creatures is preserved through the winter.

Another property of seawater relevant to climate is that its freezing temperature is below 0°C. As shown in Figure 5.2b, and by the gray line in Figure 5.1, the freezing point of water with a salinity of 30 psu is –1.8°C and a 5 psu increase in salinity corresponds to a decrease of 0.28°C in the freezing temperature.

When ocean water begins to freeze and sea ice forms, salt is rejected from the ice crystal structure, increasing the salinity of the water below the sea ice and decreasing its freezing temperature. The fresher (lighter) sea ice floating on

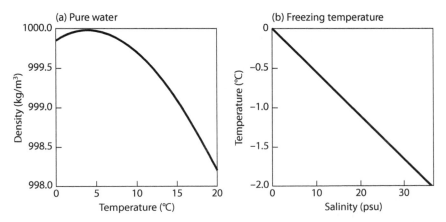

Figure 5.2 Relationships between (a) temperature and density for freshwater and (b) salinity and freezing temperature for seawater.

the saltier (more dense) seawater blocks the transfer of heat between the ocean and the atmosphere. With a sea ice barrier in place, atmospheric temperatures up to 30°C cooler than sub-ice ocean temperatures have been observed.

5.2 THE FIRST LAW OF THERMODYNAMICS

The first law of thermodynamics is a statement of conservation of energy. Heating (or cooling) of a parcel of air or water is balanced when the parcel changes temperature, does work on the environment, or both. One form of the thermodynamic equation is

$$c_v \frac{dT}{dt} + p \frac{d\alpha}{dt} = Q, \tag{5.2}$$

where Q is the rate at which heat is added to a parcel of air or water (W/m^2), called the *diabatic heating rate*. Radiative heating is one component of the diabatic heating. Another form is sensible heating, due to conduction. Latent heat releases when water condenses, and cooling due to evaporation, also contribute to the diabatic heating rate.

The first term on the left in Eq. 5.2 represents a change in temperature of the parcel if the volume is fixed, that is, if the parcel is constrained not to do work on the environment by expanding. The constant c_v is the specific heat at constant volume, essentially a proportionality constant that translates between the diabatic heating rate (Q) and the rate of temperature change (dT/dt). The second term on the left in Eq. 5.2 represents work done by the parcel in increasing its volume in response to the addition of heat (a negative value means that work is being done on the parcel by the environment when a parcel contracts). Here, the symbol α denotes the specific volume, or the volume occupied by a unit mass: $\alpha \equiv 1/\rho$.

Another form of the thermodynamic equation is also useful for atmospheric applications. Because changes in density are difficult to observe, the ideal gas law is used obtain

$$c_p \frac{dT}{dt} - \alpha \frac{dp}{dt} = Q, \tag{5.3}$$

where $c_p = 1004 \; J/(kg \cdot K)$ is the specific heat of air with pressure held constant. (See exercise 5.1.)

Under *adiabatic conditions*, that is, $Q = 0$, Eq. 5.3 can be written as

$$\frac{dT}{T} - \frac{R}{c_p} \frac{dp}{p} = 0 \quad \text{or} \quad d\ln T = \frac{R}{c_p} d\ln p, \tag{5.4}$$

where the ideal gas law (Eq. 5.1) was used to eliminate α. Integrating from the surface (T_s, p_s) to some level in the atmosphere (T, p) gives

$$T_S = T \left(\frac{p_S}{p} \right)^{R/c_p}. \tag{5.5}$$

This is one of Poisson's equations, which express the constraints on changes in the three state variables of an ideal gas under adiabatic conditions.

Consider a parcel of air with temperature T_S originating at the surface. Equation 5.5 indicates how the temperature of that parcel will change as it moves around in the atmosphere experiencing various values of air pressure, p, away from the surface. (As an air parcel moves, its pressure equilibrates with the environmental air pressure very quickly.) According to Eq. 5.3, with $Q = 0$, the parcel expands and cools or contracts and warms as pressure changes, even though no heat is added to the parcel. These temperature changes are *adiabatic*. If the parcel travels back to the surface, it recovers its original temperature, T_S. In other words, T_S is *conserved* as the parcel moves adiabatically in the atmosphere.

Based on this idea of a conserved temperature under adiabatic conditions, potential temperature, Θ, is defined as the temperature a parcel would have if brought adiabatically to the surface pressure, p_S:

$$\Theta \equiv T \left(\frac{p_S}{p} \right)^{R/c_p}. \tag{5.6}$$

In the absence of information about surface pressure, p_s in Eq. 5.6 is often replaced by a reference pressure, p_0, which can be set equal to the globally averaged surface pressure of 1013 hPa, or simply to 1000 hPa. The adiabatic thermodynamic equation can then be written

$$\frac{d\Theta}{dt} = 0. \tag{5.7}$$

5.3 HEAT BALANCE EQUATIONS

Constructing heat balance equations allows us to generate expressions that can be used to understand the flow of heat through the climate system. In an equilibrium heat balance equation, heat input equals heat output and there is no temperature trend. If a system (or subsystem) is not in thermal equilibrium, a temperature trend will occur.

EQUILIBRIUM HEAT BALANCES

At the top of the atmosphere, only radiative processes need be considered in the heat balance. Energy enters the climate system in the form of shortwave radiation and the earth system emits longwave radiation. If the earth system is assumed to be in radiative equilibrium with the sun, the heat balance at the top of the atmosphere is simply

$$\frac{S_0(1-\alpha)}{4} = \varepsilon \sigma T_E^4, \tag{5.8}$$

as discussed in section 4.2.

At the surface of the earth, other forms of heat and other processes of heat transfer in addition to radiation must be considered. The exchange of heat between the atmosphere and the surface is expressed in terms of three forms of energy. One is radiative, including both longwave and shortwave radiative fluxes. The other two are sensible and latent heat fluxes, denoted here by H_S and H_L, respectively, and often referred to as the *turbulent heat fluxes*. For the thermal equilibrium case, and in the absence of vertical or horizontal heat fluxes within the surface, the surface heat balance is

$$(1-\alpha_S)S_{INC} - F_{NET} - H_S - H_L = 0. \qquad (5.9)$$

Each term in Eq. 5.9 is discussed in further detail.

(1) S_{INC}, the rate at which shortwave (solar) radiation is incident on the surface, was defined in Eq. 4.8, and α_S is the albedo of the surface material. Therefore, $(1-\alpha_S)S_{INC}$ is the rate at which shortwave radiation is absorbed by the surface, analogous to Eq. 4.9, which considered the rate at which the incident solar radiation is absorbed by the entire earth system. Surface albedos vary widely, from values of about 0.1 over water to values greater than 0.8 for fresh snow. Table 5.1 lists typical albedo values for some common surfaces.

(2) The net longwave radiative heating of the surface, F_{NET}, is the difference between the upward emission from the surface and the downward back radiation from the atmosphere:

$$F_{NET} = \varepsilon\sigma T_S^4 - F_{BACK}. \qquad (5.10)$$

Here, the emissivity of the surface material has been accounted for (see Eq. 4.2). Longwave emissivities for common surface substances tend to be close to 1 (Table 5.1), but the values depend on wavelength as well as on other factors such as the moisture content of the surface.

(3) The surface sensible heat flux, H_S, is the rate at which heat is transferred between the atmosphere and the surface by *conduction*. Heat conduction is defined as the transfer of heat down a temperature gradient through the interactions and random motions of adjacent molecules. However, in the field of climate dynamics, sensible heating also includes the transport of heat by incoherent motion. The term is also used to encompass the effects of *turbulence* on heat transport, which refers to small-scale irregular motion as well as to small-scale organized circulations such as thermals.

The sensible heat content of a unit mass of air at temperature T is $c_p T$, where c_p is the specific heat at constant pressure, or the amount of heat it takes to raise the temperature of 1 kg of air by 1 K [$c_p = 1004$ J/(K · kg)] without allowing the parcel to do work. Thus, $c_p T$ represents the amount of energy (number of joules) that was "invested" to bring each kilogram of a parcel to its observed temperature. The sensible heat content of a unit volume of air is $\rho c_p T$.

A *heat flux* is the amount of heat crossing a unit surface area per unit time, expressed in units of J/(m² · s) or, equivalently, W/m². In formulating a mathematical expression for H_S, imagine that a unit volume of air moves from the surface to a height of 1 m in 1 second, so the vertical velocity is $w = 1$ m/s. In this case, an amount of sensible heat given by $\rho c_p T$ is removed from the surface

Table 5.1. Representative shortwave albedos and longwave emissivities relevant to the climate system

Material		Shortwave albedo	Longwave emissivity
Water[a]		0.08	0.95–1.00
Snow	Fresh	0.70–0.90	0.82
	Old	0.45–0.60	0.89
Ice		0.35	0.98
Tundra		0.20	0.95–0.97
Desert		0.38	0.95–0.96
Tropical savanna		0.18	0.97–0.99
Grassland		0.10–0.20	0.97–0.99
Cropland		0.15–0.25	0.97–0.99
Tropical forest		0.08–0.13	1.00
Boreal forest		0.16	0.98
Soil	Wet sand	0.25	0.90
	Dry sand	0.40	0.76
	Wet soil	0.06–0.13	0.66
	Dry soil	0.22–0.33	0.94
Clouds	Cumulus	0.65–0.75	0.80[b]
	Cirrus	0.45–0.60	0.10[b]
	Stratus	0.35–0.55	0.99[b]
Manufactured material	Asphalt	0.10	0.93
	Concrete	0.40	0.96
	Brick	0.30	0.93

[a] Assumes that the sun is directly overhead.

[b] Values are approximate because cloud emissivities vary greatly depending on the cloud's composition (e.g., liquid water and ice content, droplet sizes) and wavelength.

(or contact with the surface) into the atmosphere across a unit area in 1 second. Then, the sensible heat flux is

$$H_S = w\rho c_p T. \tag{5.11}$$

(4) The *latent heat flux*, H_L, is the rate at which heat is exchanged between the atmosphere and the surface due to liquid-to-vapor (and vice versa) phase

changes of water. The latent heat flux is positive, that is, heat is transferred from the surface to the atmosphere, when water evaporates from the surface. A negative latent heat flux indicates that heat is transferred from the atmosphere to the surface by the condensation of water onto the surface, for example, when dew forms.

Consider a 1-kg parcel of air that contains water vapor. (The 1-kg mass of the parcel consists of dry air plus water vapor.) The amount of energy that was used to evaporate the water contained in the 1-kg parcel is Lq, where L is the latent heat of vaporization (the amount of energy needed to evaporate 1 kg of water) and q is the specific humidity (section 2.3). The value of L (see Appendix A) is temperature dependent. For water at $0°C$, $L = 2.5 \times 10^6$ J/kg-H_2O. Multiply by the density of the moist air parcel to find the energy required to evaporate the water contained in a parcel of unit volume. Note the units:

$$\rho L q \sim \left(\frac{\text{mass of parcel}}{\text{m}^3}\right)\left(\frac{\text{J}}{\text{kg-}H_2O}\right)\left(\frac{\text{kg-}H_2O}{\text{mass of parcel}}\right) \sim \frac{\text{J}}{\text{m}^3}.$$

Therefore, the expression for the latent heat flux analogous to Eq. 5.11 is

$$H_L = w\rho L q. \tag{5.12}$$

To develop a more physical understanding about how turbulent fluxes work at the atmosphere/surface interface, divide each dependent variable in Eqs. 5.11 and 5.12 into time-mean and time-varying components, denoted by overbars and primes, respectively:

$$w(\lambda,\phi,z,t) = \overline{w}(\lambda,\phi,z) + w'(\lambda,\phi,z,t);$$
$$T(\lambda,\phi,z,t) = \overline{T}(\lambda,\phi,z) + T'(\lambda,\phi,z,t); \tag{5.13}$$
$$q(\lambda,\phi,z,t) = \overline{q}(\lambda,\phi,z) + q'(\lambda,\phi,z,t).$$

The time-varying component is called the (temporal) perturbation, and the time average of the perturbation quantities is zero. This is shown mathematically by taking the time mean of Eqs. 5.13. For example, since the time mean of $\overline{w}$, denoted $\overline{\overline{w}}$, is just $\overline{w}$,

$$\overline{w(\lambda,\phi,z,t)} = \overline{\overline{w}(\lambda,\phi,z) + w'(\lambda,\phi,z,t)} = \overline{\overline{w}(\lambda,\phi,z)} + \overline{w'(\lambda,\phi,z,t)}$$
$$\Rightarrow \overline{w'(\lambda,\phi,z,t)} = 0. \tag{5.14}$$

Equivalently, in integral form for a time-averaging period P,

$$\overline{w(\lambda,\phi,z,t)} \equiv \frac{1}{P}\int_0^P w(\lambda,\phi,z,t)\,dt = \frac{1}{P}\int_0^P \overline{w}(\lambda,\phi,z)\,dt + \frac{1}{P}\int_0^P w'(\lambda,\phi,z,t)\,dt$$

$$= \overline{w}(\lambda,\phi,z) + \frac{1}{P}\int_0^P w'(\lambda,\phi,z,t)\,dt \tag{5.15}$$

$$\Rightarrow \frac{1}{P}\int_0^P w'(\lambda,\phi,z,t)\,dt = \overline{w'(\lambda,\phi,z,t)} = 0.$$

Substituting Eqs. 5.13 into Eqs. 5.11 and 5.12 and then taking the time mean we get

$$\overline{H}_S = \rho c_p\left[\overline{\overline{w}\,\overline{T}} + \overline{w'\,\overline{T}} + \overline{\overline{w}T'} + \overline{w'T'}\right] = \rho c_p\left[\overline{w}\,\overline{T} + \overline{w'T'}\right] \qquad (5.16)$$

and

$$\overline{H}_L = \rho L\left[\overline{\overline{w}\,\overline{q}} + \overline{w'\,\overline{q}} + \overline{\overline{w}q'} + \overline{w'q'}\right] = \rho c_p\left[\overline{w}\,\overline{q} + \overline{w'q'}\right] \qquad (5.17)$$

where ρ is assumed to be constant in time. Close to the surface, where $\overline{w} \cong 0$,

$$\overline{H}_S = \rho c_p\overline{w'T'}, \qquad (5.18)$$

and

$$\overline{H}_L = \rho L\overline{w'q'}. \qquad (5.19)$$

According to Eq. 5.18, sensible heat is transferred vertically in the atmosphere when w' and T' are correlated. When $w' > 0$ and $T' > 0$, upward perturbation velocities are correlated with warm temperature anomalies, so warm air is rising. In this case, $\overline{w'T'} > 0$, and there is a positive (upward) sensible heat flux from the surface to the atmosphere. When $w' < 0$ and $T' < 0$, cool air is sinking. This is also a case in which $\overline{H}_S > 0$. If cold air is rising ($w' > 0$ and $T' < 0$) or warm air is sinking ($w' < 0$ and $T' > 0$), then $\overline{H}_S < 0$. Similarly, according to Eq. 5.19, rising moist air carries latent heat from the surface into the atmosphere.

Observations show that the time scales over which w' and T', and w' and q', are correlated are on the order of seconds and that they vary strongly on small space scales as well. So while Eqs. 5.18 and 5.19 are accurate expressions, they are not practical because values of w', T', and q' cannot be measured on small enough space and time scales over the globe to provide even estimates of the turbulent fluxes.

As a replacement for the exact expressions for the sensible and latent heat fluxes, *physical parameterizations* have been developed. These parameterizations are based on the underlying physics of the turbulent fluxes and they relate H_L and H_S to commonly measured large-scale quantities. For example, the *bulk aerodynamic formulas* are used to represent sensible and latent heat fluxes from the surface in computer models and observational analyses:

$$H_S = \rho c_p C_{DH} V(T_S - T_A) \qquad (5.20)$$

$$H_L = \rho L C_{DL} V(q_S - q_A) \qquad (5.21)$$

where C_{DH} and C_{DL} are *drag coefficients*, often taken to be 0.001 over water and 0.003 over land. V is the surface horizontal wind speed, T_S is the surface temperature, q_A is the specific humidity of the surface air, and q_S is the saturation specific humidity at temperature T_S. Note that the density of air, ρ, is used in these parameterizations, not the density of the surface material.

NONEQUILIBRIUM SURFACE HEAT BALANCE

Equation 5.9 assumes the net heating of the surface is zero and that there are no horizontal and vertical fluxes of heat into and out of the surface material. A more general surface heat balance equation allows for these processes.

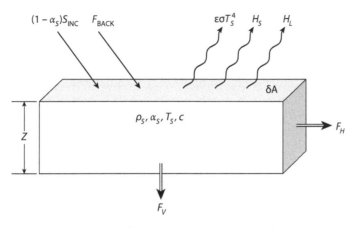

Figure 5.3 A volume, V, of surface material exchanging heat with the atmosphere as well as with adjacent surface material.

Consider the volume $\delta V = Z\delta A$ of surface material (land or water) drawn in Figure 5.3. The surface of the volume exchanges heat with the atmosphere across a unit surface area δA. This surface has an albedo α_S. The density of the surface material is ρ_S, and it is assumed to be isothermal at temperature T_S. The depth of the volume, Z, can be interpreted as the mixed-layer depth for a heat balance calculation over the ocean or the depth to which heat penetrates the soil over land.

If the components of the heat budget do not balance, the surface will heat or cool at a rate $\partial T_S/\partial t$ determined by

$$C\frac{\partial T_S}{\partial t} = (1-\alpha_S)S_{INC} + F_{BACK} - \varepsilon\sigma T_S^4 - H_S - H_L - F_H - F_V \qquad (5.22)$$

where C is the *heat capacity* of the surface material, defined as the amount of heat required to raise the temperature of the volume by 1 K. F_H and F_V are the net horizontal and downward heat fluxes out of the volume, respectively. For example, if the volume is in the ocean F_H captures the effects of horizontal ocean currents transporting energy out of (or into) a volume of water. The term F_V can account for the effects of downward diffusion of heat within soil or water, and upwelling and downwelling in the ocean.

The value of the heat capacity, C, depends on two factors. One is the specific heat capacity, c, usually called just the specific heat, of the surface material. This is the amount of heat required to raise the temperature of 1 kg of the surface material by 1 K. (See Appendix A for specific heats of some substances relevant to the climate system.)

The other factor that determines C is the mass of material associated with a unit surface area, that is, the mass heated by the net energy absorbed across the surface area δA. Denoted by M_A, with units of kg/m^2, this mass is

$$M_A = \frac{\text{mass of the volume}}{\text{surface area of the volume}} = \frac{\rho_S V}{\delta A} = \rho_S Z. \qquad (5.23)$$

Then,

$$C = cM_A = c\rho_s Z. \tag{5.24}$$

The nonequilibrium surface heat balance can be used to understand the physical basis of continentality. Suppose there is a net heating of the surface of 20 W/m^2, so Eq. 5.22 is

$$C\frac{\partial T_s}{\partial t} = 20 \text{ W/m}^2. \tag{5.25}$$

If this net heating is incident on the ocean, in a region where the mixed-layer depth is 100 m, then

$$C = c_w\rho_w Z = [4218 \text{ J/(kg} \cdot \text{K)}](10^3 \text{ kg/m}^3)(100 \text{ m}) = 4.2\times10^8 \text{J/(m}^2\cdot\text{K)}. \tag{5.26}$$

Rearranging (5.25), we can estimate the time required to increase the temperature of the ocean mixed layer by 2 K as

$$\Delta t = \frac{C\Delta T_s}{20\,\text{W/m}^2} \approx 1167 \text{ hr} = 48 \text{ days}. \tag{5.27}$$

If the net heating is incident over land, say, onto a dry sand surface with a heat penetration depth of 2 cm, then

$$C = c_s\rho_s Z = [840 \text{ J/(kg}\cdot\text{K)}](2.65\times10^3\text{kg/m}^3)(2\times10^{-2}\text{m}) = 4.5\times10^4\text{J/(m}^2\cdot\text{K)}, \tag{5.28}$$

and $\Delta t = 1.25$ hr. Thus, the land surface responds to the net surface heating on diurnal time scales, while the ocean responds on seasonal time scales, as indicated in Table 3.1.

5.4   OBSERVED HEAT FLUXES

GLOBAL HEAT BALANCE CLIMATOLOGY

Figure 5.4 summarizes the flow of heat into, out of, and within the climate system in the global and annual average. The numbers (in W/m^2) are rough estimates based on satellite observations (see Additional Reading), and they may not be exactly the same as other estimates. The figure provides only a broad overview of the global heat balance.

The left side of Figure 5.4 outlines the fate of solar energy as it passes through the atmosphere. Solar radiation of 341 W/m^2 is incident at the top of the atmosphere in this estimate with 102 W/m^2 reflected back to space from the earth system, implying that the average planetary albedo is 0.3. Much of this reflection is from clouds (~62 W/m^2), but backscattering from molecules and reflection from the surface also contribute (~17 W/m^2 and ~23 W/m^2, respectively). The remaining solar radiation, about 239 W/m^2, fuels the climate system. Roughly 67%, or 161 W/m^2, of this energy passes through the atmosphere and is absorbed by the surface. Some of this solar radiation arrives at the surface in

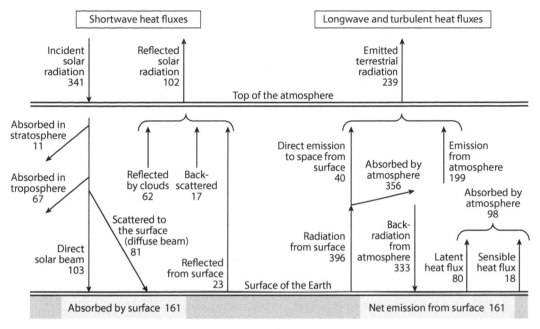

Figure 5.4 The estimated global heat balance (W/m²). Based on Trenberth, Fasullo, and Kiehl (2009).

the direct solar beam (~103 W/m²) and some in the diffuse beam (~81 W/m²) after multiple Rayleigh and Mie scattering events in the atmosphere. Only a relatively small amount of the total incoming solar radiation (~11 W/m²) is absorbed by gases (e.g., ozone) in the stratosphere, and about 67 W/m² is absorbed by aerosols, water vapor, and water droplets in the troposphere.

The right half of Figure 5.4 illustrates the passage of outgoing longwave, or terrestrial, radiation through the atmosphere. The longwave emission rate from the surface (~396 W/m²) implies a surface temperature of about 289 K according to Eq. 4.2 with $\varepsilon = 1$. Only about 10%, or ~40 W/m², of this emission from the surface radiates directly to space, largely through the longwave atmospheric window (8–13 μm).

The lion's share of the longwave emission from the surface (~356 W/m²) is absorbed within the atmosphere. This is the primary heat source for the atmosphere, so the atmosphere is heated from below (from the surface) and not from above (directly by the incident solar radiation). Molecules, water droplets, clouds, and dust in the atmosphere reradiate the absorbed longwave energy at their own temperatures. About 333 W/m² is radiated back to the surface where it is absorbed by the land or ocean. This is the longwave back radiation associated with the greenhouse effect (section 4.5). Outgoing longwave radiation (OLR) includes emission from clouds (~30 W/m²) and atmospheric molecules (~169 W/m²). These combine with the 40 W/m² emitted directly from the surface to space, balancing the 239 W/m² solar absorbed.

Longwave emission from the surface (~396 W/m²) is more than twice the amount of shortwave radiation absorbed (~161 W/m²), and the largest source of heat to the surface is the longwave back radiation from the atmosphere ($\sim$333 W/m^2). The surface and the lower atmosphere are tightly coupled by this vigorous exchange of longwave radiation.

The *net* longwave cooling of the surface is 63 W/m^2. The latent heat flux from the surface is about 80 W/m^2, while the sensible heat flux is estimated at 18 W/m^2. The *Bowen ratio*, defined as the ratio of the sensible and latent heat fluxes, is about 0.23 in the global mean, but the relative values of the turbulent heat fluxes vary widely from place to place.

In the estimated global heat budget shown in Figure 5.4, the net shortwave heating of the surface (161 W/m^2) is balanced by the net cooling of the surface (also 161 W/m^2). The approximated fluxes at the top of the atmosphere balance as well. This heat budget represents a system in balance, with no temperature trend. Changes in the atmosphere's composition perturb this balance. Increases in atmospheric CO_2 and other greenhouse gases increase the longwave back radiation and change the balance at the surface. If snow and ice coverage is reduced as a consequence, the solar radiation absorbed will increase, and there will be additional repercussions throughout the system. In chapter 11, a simple climate model based on the surface heat balance is used to estimate temperature changes due to increasing greenhouse gases.

OBSERVED DISTRIBUTION: TOP OF THE ATMOSPHERE

The heat balance at the top of the atmosphere has only two components, namely, the shortwave absorbed by the system and the OLR (Eq. 5.8). If the climate system is in an equilibrium state, these components must balance in the global and climatological mean, but they need not balance locally and, in fact, they do not.

The annual mean distribution of solar radiation incident at the top of the atmosphere (Fig. 5.5a) is zonally uniform and symmetric about the equator.[1] It decreases by a factor of about 2 from the equator to 70° latitude due to the curvature of the earth.

The solar radiation absorbed by the earth system (Fig. 5.5b) has east–west variations in addition to meridional structure. It varies due to variations in the planetary albedo (shown in Fig. 5.6) that are associated primarily with cloud and surface albedo distributions. Note that while the incident solar radiation varies by a factor of about 2 between the equator and high latitudes, the solar radiation absorbed changes by a factor of 3 or 4. The reason for this difference is the latitudinal dependence of the planetary albedo.

Two factors cause the strong increase in planetary albedo with latitude. One is that surface albedos are larger at high latitudes due to the presence of snow and ice on the surface (see Table 5.1). Another important factor is the dependence of the albedo of water on the solar zenith angle. When the solar zenith angle is large, the albedo of water is much larger than the nominal value of

---

[1] This figure, and the figures of heating components that follow, are from the ERA-Interim reanalysis climatology for 1979–2009. This reanalysis is provided by the European Center for Medium Range Weather Prediction. The full reference is given in chapter 2.

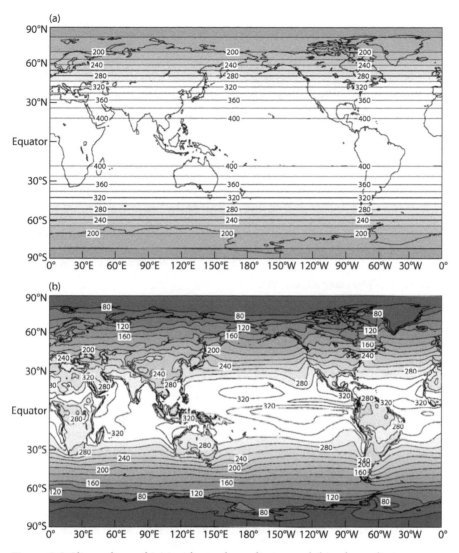

Figure 5.5 Climatology of (a) incident solar radiation and (b) solar radiation absorbed at the top of the atmosphere. Contour intervals are 20 W/m².

0.08 which is accurate only for solar zenith angles of less than about 40°. At higher zenith angles, the albedo of water drops significantly as illustrated in Figure 5.7. In the late afternoon or early morning, many of us have experienced a strong glare from water standing on a road, or from the open water of a lake near sunrise or sunset. This albedo effect is most pronounced under clear-sky conditions, when much of the solar radiation remains in the direct solar beam. Since the annually averaged zenith angle is smaller close to the equator, a higher fraction of the incident solar radiation is absorbed at low latitudes. This effect causes the latitudinal gradients in the planetary albedo over the oceans in middle latitudes (see Fig. 5.6).

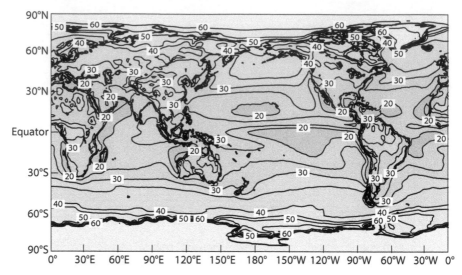

Figure 5.6 The planetary albedo (%).

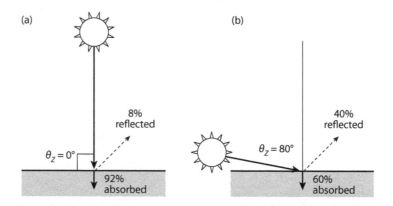

Figure 5.7 The dependence of water albedo on solar zenith angle, θ_z.

Figure 5.8 shows the annually averaged longwave radiative flux emitted by the earth system, that is, the OLR. Large values of OLR are generated at low latitudes where emission temperatures are warmer, and in regions with low clouds or few clouds where the OLR originates closer to the surface (and, therefore, at higher temperatures).

A comparison of Figures 5.5b and 5.8 shows that the solar radiation absorbed at tropical latitudes is greater than the longwave radiation emitted. The opposite is true at high latitudes. Figure 5.9a shows the difference—the solar radiation absorbed minus the longwave radiation emitted at the top of the atmosphere. The net radiative heating is positive in the tropics where the solar radiation absorbed is up to 60 W/m² greater than the OLR. It is negative at higher latitudes, where the OLR exceeds the solar radiation absorbed by up to 120 W/m². The latitude at which the solar radiation absorbed and the OLR balance is close to 30° in both hemispheres.

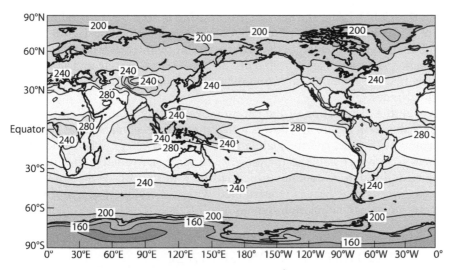

Figure 5.8 The annual mean OLR climatology (W/m²).

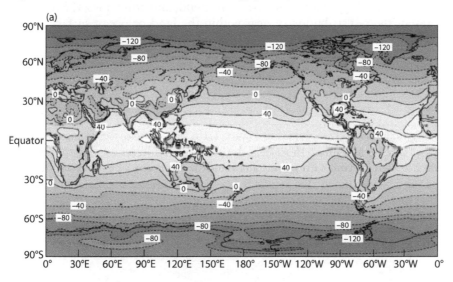

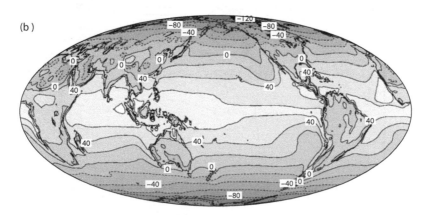

Figure 5.9 Net radiation at the top of the atmosphere in (a) equidistant cylindrical and (b) Mollweide projections. Contour intervals are 20 W/m².

In the global mean, the net radiative heating is zero for an equilibrium climate (Eq. 5.8), but it is hard to believe that when examining Figure 5.9a because the projection enlarges the surface area at high latitudes. The *Mollweide projection* shown in Figure 5.9b plots surface area accurately and shows more clearly that the *radiation surplus* in the tropics balances the *radiation deficit* in high latitudes.

Given the distribution of the net radiative heating at the top of the atmosphere shown in Figure 5.9, how is it that the tropical climate is not constantly warming with time, and the high latitude climate cooling? Why is the climate system (relatively) stable in the presence of large net radiative heating in the tropics and radiative cooling at high latitudes at the top of the atmosphere? The answer is that heat is redistributed within the climate system by atmospheric and ocean circulation systems.

OBSERVED DISTRIBUTION: SURFACE

For the climatology, with no temperature trend, and combining the terms that represent the redistribution of heat within the land or ocean surface, we can rewrite Eq. 5.22 as

$$0 = S_{ABS} + F_{BACK} - \varepsilon \sigma T_S^4 - H_S - H_L - F_{SFC} \tag{5.29}$$

where

$$S_{ABS} = (1 - \alpha_S) S_{INC} \text{ and } F_{SFC} = F_H + F_V.$$

The S_{ABS} climatology is plotted in Figure 5.10. Values near the equator are three to four times greater than those near the poles. This difference is partly due to the decrease with latitude of the solar flux incident at the top of the atmosphere (Fig. 5.5a) and partly due to albedo effects associated with the angle of incidence (Fig.

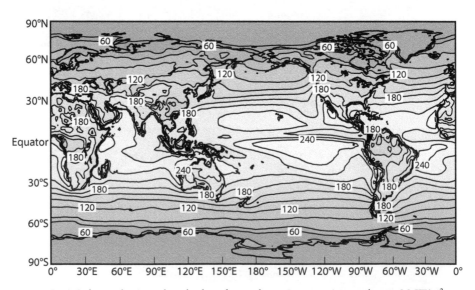

Figure 5.10 Solar radiation absorbed at the surface. Contour intervals are 20 W/m².

5.7). Additionally, surface snow and ice become relevant at high latitudes. There is a minimum close to the equator due to the deep cloud cover associated with the ITCZ, and higher values in the subtropics where the sky is relatively clear.

The two longwave radiation components of the surface heat balance are drawn in Figure 5.11. Terrestrial emission (Fig. 5.11a) is proportional to the fourth power of the surface temperature (Eq. 4.2), so its distribution is similar to the low-level and surface temperature distributions shown in Figures 2.6 and 2.15. The longwave back radiation (Fig. 5.11b) is larger at low latitudes because of the distribution of atmospheric water vapor (Fig. 2.31), which is a powerful greenhouse gas that occurs in higher concentrations in warmer regions (see Figs. 2.30–2.32).

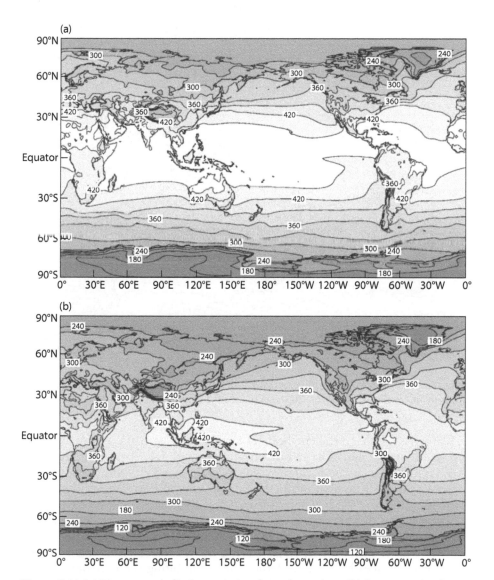

Figure 5.11 (a) Longwave radiative emission from the surface. (b) Longwave back radiation from the atmosphere to the surface. Contour intervals are 30 W/m².

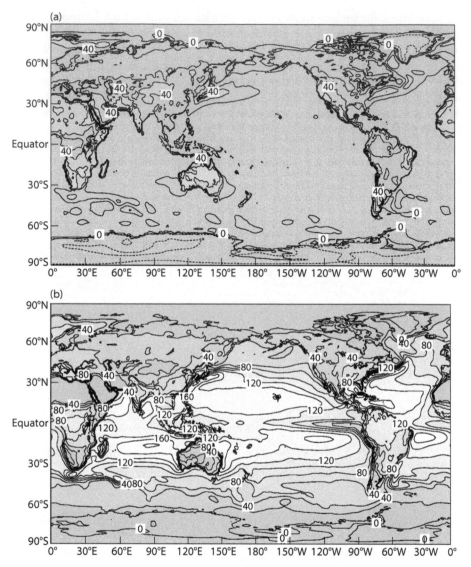

Figure 5.12 (a) Sensible and (b) latent heat fluxes from the surface to the atmosphere. Contour intervals are 20 W/m². Dashed contours indicate negative values.

Estimates of the turbulent heat flux climatologies from a reanalysis are shown in Figure 5.12. The sensible heat flux (Fig. 5.12a) is generally larger over land than over the oceans, except over the western boundary currents, for example, the Kuroshio in the Pacific and the Gulf Stream in the Atlantic (see Figs. 2.15 and 2.22) where warm water underlies cooler air (Eq. 5.20). Negative values in Antarctica, the Arctic, and Greenland indicate a downward (from atmosphere to surface) sensible heat flux, since the overlying air is often warmer than the surface, especially during the winter. Ice and water on land and ocean surfaces have an insulating effect, inhibiting the exchange of heat between the atmosphere and ocean and suppressing the sensible heat flux.

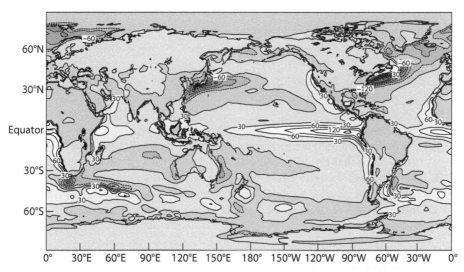

Figure 5.13 Contribution of horizontal and vertical heat fluxes within the surface to the local heat balance. Contour interval is 30 W/m². Dashed contours indicate negative values.

In contrast to the sensible heat flux, the latent heat flux is larger over the oceans than over land due to the essentially limitless supply of water for evaporation. The latent heat flux is the product of the evaporation rate and the latent heat of vaporization so the patterns of Fig. 2.29 are repeated in Figure 5.12b, but more detail is displayed in the latter. Maximum latent heat fluxes occur in the subtropics where the surface is warm, trade winds provide a strong, steady surface wind at all times of the year, and the overlying air is dry.

Finally, Figure 5.13 displays heat transport within the surface, F_{SFC}. (This term is calculated as a residual using reanalysis values in Eq. 5.29.) When $F_{SFC} > 0$, ocean currents and/or upwelling are contributing to regional cooling of the surface temperature. F_{SFC} is small over land, since the solid surface does not conduct much heat horizontally or even vertically. The large positive values in the tropical east Pacific and Atlantic Oceans indicate that the eastern boundary currents and upwelling contribute to the lower temperatures of the cold-tongue regions. Negative values in the vicinity of the western boundary currents—the Gulf Stream in the western North Atlantic and the Kuroshio in the western North Pacific—signify the warming role of these currents in middle and high latitudes.

5.5 ADDITIONAL READING

A contemporary estimate of the global heat balance, including a discussion of uncertainties, is provided in the article "Earth's Global Energy Budget" by K. E. Trenberth, J. T. Fasullo, and J. Kiehl in the *Bulletin of the American Meteorological Society* 90:311 (2009). Fig. 5.4 is based on this estimate.

5.6 EXERCISES

5.1.

(a) Derive Eq. 5.3 from Eq. 5.2. Start by taking the total time derivative of the idea gas law, and then use Eq. 5.2 to eliminate α. Note that $c_p = c_v + R$.

(b) Derive Eq. 5.5.

5.2. Show that potential temperature is conserved under adiabatic conditions. *Hint*: Show that Eq. 5.7 is equivalent to Eq. 5.3 with $Q = 0$.

5.3. Is the planetary albedo equal to the sum of the surface albedo and the atmospheric albedo? Explain your answer, perhaps with a diagram.

5.4. Instantaneous measurements over a layer of water 5 cm thick are as follows:

incident solar radiation $= 250$ W/m^2
water temperature $= 287$ K
atmospheric back radiation $= 275$ W/m^2
sensible heat flux $= 30$ W/m^2
latent heat flux $= 70$ W/m^2

The sun is overhead. Is the water heating up or cooling down? At what rate is its temperature changing? After 10 minutes, do you expect that the rate of change of temperature of the water will be larger or smaller? Explain.

5.5. Comparison of thermal emission and the sensible heat flux at the earth's surface.

(a) Use a Taylor series expansion to show that the blackbody emission from a surface can be written

$$F = \sigma T_s^4 = \sigma T_0^4 + 4\sigma T_0^3 (T_s - T_0) + \ldots,$$

where T_0 is some reference temperature. (Write out at least four terms on the right-hand side of the above equation.)

(b) Show that if T_0 and T_s are only a few degrees apart, that is, if you choose the reference temperature T_0 to be close to the surface temperature T_s, then retaining only two terms in the Taylor expansion is a reasonable approximation.

(c) Write the bulk aerodynamic formula for the sensible heat flux. Compare the form of this equation with your approximation for blackbody emission. (For example, what if you choose $T_0 = T_A$ in the blackbody emission temperature formula?)

(d) Derive equations that express the sensitivity of longwave and sensible heat fluxes to changes in surface temperature. [*Hint*: The sensitivity of a flux F to T_s is expressed mathematically as $\partial F / \partial T_s$.]

(e) Which is more sensitive to changes in surface temperature—longwave or sensible heat fluxes?

DYNAMICS: THE FORCES THAT DRIVE ATMOSPHERIC AND OCEAN CIRCULATIONS

In chapter 5 we showed that the longwave and shortwave radiative heat fluxes at the top of the atmosphere do not balance locally—there is net radiative heating in the tropics and radiative cooling at high latitudes (see Fig. 5.9). Why, then, are temperatures in the tropics not continually increasing and those at high latitudes decreasing? The reason is that atmospheric and ocean circulation systems move heat from the tropics to higher latitudes to balance local radiative heating and cooling.

To understand these circulation systems, the forces that drive and constrain them can be accounted for by applying Newton's second law,

$$\sum \vec{F} = m\vec{a}, \qquad (6.1)$$

to a parcel of air or ocean water. (A *parcel* is a unit of air or ocean water, defined either as a unit mass or a unit volume.) In Eq. 6.1, the summation indicates that various forces act on parcels to determine acceleration. That acceleration, of course, generates velocity—wind or ocean currents—which changes according to

$$\frac{d\vec{v}}{dt} = \frac{\sum \vec{F}}{m}. \qquad (6.2)$$

The use of the total, or Lagrangian, derivative in Eq. 6.2 (see Appendix C) indicates that the perspective is one of following the parcel, much as Newton's first law of motion is often illustrated in physics textbooks by adding the forces acting on a block as it slides down an inclined plane. The translation from the Lagrangian to the Eulerian perspective, which deals with motion relative to a given location in a coordinate system, is discussed below and in Appendix C. First, the most important forces that govern large-scale motion in the atmosphere and oceans are assembled for the summation in Eq. 6.2.

The scalar components of the vector equation, Eq. 6.2, are known as the *equations of motion* or the *momentum equations*. Using the local Cartesian coordinate system defined in Appendix B, we have

$$\frac{du}{dt} = \frac{\sum F^x}{m} = F^x_{COR} + F^x_{PGF} + F^x_F, \qquad (6.3)$$

$$\frac{dv}{dt} = \frac{\sum F^y}{m} = F^y_{COR} + F^y_{PGF} + F^y_F, \qquad (6.4)$$

and

$$\frac{dw}{dt} = \frac{\sum F^z}{m} = F^z_{COR} + F^z_{PGF} + F^z_{GRAV} + F^z_{F}, \qquad (6.5)$$

where the most important forces governing large-scale motion in the atmosphere and oceans are included, namely, the Coriolis force per unit mass (F_{COR}), the pressure gradient force per unit mass (F_{PGF}), gravitation (F_{GRAV})and friction (F_{F}). Each of these is discussed below, where we refer to, for example, the "Coriolis force per unit mass" simply as the "Coriolis force."

6.1 THE CORIOLIS FORCE

The Coriolis force is an "apparent force" that arises because of the rotating frame of reference from which we observe the atmosphere and oceans. The laws of Newtonian physics apply in an *absolute, or inertial, frame of reference*, but we observe the atmosphere and oceans within the *noninertial frame of reference* rotating with the earth. With the Coriolis force included, the governing equations in the rotating frame of reference (seem to) conserve angular momentum, and centrifugal accelerations are accounted for.

CONSERVATION OF ABSOLUTE ANGULAR MOMENTUM

A parcel of air or water with constant mass m moving with velocity $\vec{v}$ in a circular orbit must conserve absolute angular momentum (angular momentum in the absolute frame of reference) in the absence of externally applied torques. Angular momentum, $\vec{L}$, is a vector quantity, defined as

$$\vec{L} \equiv \vec{r} \times m\vec{v}, \qquad (6.6)$$

where $m\vec{v}$ is linear momentum, and $\vec{r}$ is the moment arm extending from the axis of rotation to the parcel (Figure 6.1). The magnitude of $\vec{L}$ is $mv_T r$, where v_T is the tangential velocity and the direction of $\vec{L}$ is given by the *right-hand rule*. (Curl the fingers of your right hand in the direction of rotation and your

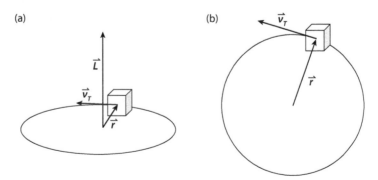

Figure 6.1. Definition of angular momentum.

thumb will point in the direction of $\bar{L}$.) Absolute angular momentum per unit mass, M, is

$$M = v_T r. \qquad (6.7)$$

Observations of atmospheric and ocean circulation are invariably presented in the rotating, noninertial frame of reference. The winds and currents in chapters 2 and 3, for example, represent velocity relative to the rotating earth, but the motion of a parcel is constrained by conservation of angular momentum in the absolute (inertial) frame of reference. For observers in the rotating frame of reference (like us), a parcel will seem to have forces acting on it when, in fact, there are no true forces acting on it; the parcel is simply conserving absolute angular momentum.

Consider a unit mass parcel of air or ocean water, diagrammed in Figure 6.2. (The size of the parcel as well as its distance above the earth's surface, z, are drawn wildly out of scale.) If the parcel has no velocity relative to the rotating earth, then $u = v = w = 0$, where the local Cartesian coordinate system (Appendix B) rotating with the earth is used. In this case, the parcel is said to be in *solid body rotation* with the earth. Viewed from the absolute frame of reference, in which angular momentum must be conserved, the parcel is not at rest but traveling in a circle of radius r, completing one circuit every 24 hours. The tangential velocity of the parcel in the inertial frame of reference is

$$v_T = \frac{2\pi(a+z)\cos\phi}{24 \text{ hr}}, \qquad (6.8)$$

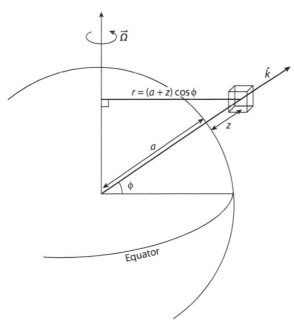

Figure 6.2. An air or ocean parcel in solid body rotation.

where the radius of the circle is the magnitude of the moment arm, r. For a parcel in the troposphere or in the ocean, $z << a$, and it is reasonable to neglect the height of the parcel above the surface of the earth compared with the radius of the earth and let

$$r = a\cos\phi. \tag{6.9}$$

This is known as the *thin atmosphere (or ocean) approximation*. Equation 6.8 indicates that the tangential velocity of a parcel in solid body rotation at the equator is 463 m/s. This is a huge speed, an order of magnitude greater than velocities observed relative to the rotating earth (Figs. 2.10–2.14).

Now, imagine that the parcel drawn in Figure 6.2 has a zonal velocity u relative to the rotating earth, that is, as measured by an observer on the ground at the same latitude as the parcel, but still $v = w = 0$. In the diagram, instantaneous westerly velocity ($u > 0$) would be motion into the page and easterly velocity ($u < 0$) would be motion out of the page. We define the following:

$\vec{\Omega}$ = angular velocity of the earth, with magnitude

$$\Omega = \frac{2\pi \text{ rad}}{24 \text{ hr}} = 7.29 \times 10^{-5} \text{s}^{-1}. \tag{6.10}$$

U_{ROT} is the instantaneous tangential velocity of the earth's surface in the absolute frame of reference at the same latitude as the parcel (i.e., the velocity of solid body rotation). Using the thin atmosphere/ocean approximation and Eq. 6.10, we have

$$U_{ROT} = \frac{2\pi r}{24 \text{ hr}} = \frac{2\pi a\cos\phi}{2\pi/\Omega} = \Omega a\cos\phi = \Omega r, \tag{6.11}$$

where the circumference of the circle with radius $2\pi a\cos\phi$ is the distance traveled in one day. Defining U_{ABS} as the instantaneous tangential velocity of the parcel in the absolute frame of reference, it is equal to the sum of the velocity of the rotating earth and the velocity relative to the rotating earth:

$$U_{ABS} = U_{ROT} + u. \tag{6.12}$$

Then, the magnitude of the absolute angular momentum per unit mass of the parcel (Eq. 6.7) is

$$M = U_{ABS}r = [\Omega(a+z)\cos\phi + u](a+z)\cos\phi \tag{6.13}$$

or, with the thin atmosphere/ocean approximation,

$$M = U_{ABS}r = (\Omega a\cos\phi + u)a\cos\phi. \tag{6.14}$$

According to Eq. 6.14, if ϕ changes, that is, if the parcel moves to the north or south, then u must also change to keep M constant (to conserve absolute angular momentum). If a parcel in the Northern Hemisphere, for example, moves to a higher latitude, then $\cos\phi$ will decrease and u must increase. This apparent zonal velocity that results from meridional motion is referred to as the *Coriolis effect*.

To evaluate the importance of the Coriolis effect, consider a parcel of air that moves from the equator to 30°N latitude due to some impulsive meridional force. In its initial position on the equator, the parcel has no zonal velocity relative to the earth. If the parcel conserves absolute angular momentum (M is constant) and moves to 30°N, the observer at 30°N will measure its zonal velocity as 134 m/s westerly (see exercise 6.2). This large zonal velocity suggests that the Coriolis effect is important. Because we do not observe values of the zonal wind comparable to 134 m/s, this comparison also suggests that other forces must be operating in the system to constrain the flow.

The equation

$$\frac{dM}{dt} = 0 \qquad (6.15)$$

expresses the principle of conservation of absolute angular momentum. (Note the Lagrangian derivative, defined in Appendix C.) Substituting Eq. 6.13 into Eq. 6.15 and carrying through the differentiation (see exercise 6.3), we obtain the zonal component of the Coriolis force,

$$\vec{F}_M = \left(2\Omega v \sin\phi + \frac{uv}{a}\tan\phi - 2\Omega w \cos\phi - \frac{uw}{a}\right)\hat{i}, \qquad (6.16)$$

where the subscript M is a reminder that this component of the Coriolis force arises due to conservation of angular momentum per unit mass in the absolute frame of reference.

CENTRIFUGAL ACCELERATIONS

The other part of the Coriolis force accounts for centrifugal accelerations that act in the rotating frame of reference and arise when a parcel's path in the absolute frame of reference is curved.

For a parcel of air or ocean water in solid body rotation with the earth ($u = v = w = 0$), the centrifugal force per unit mass, $\vec{F}_{SB}$, felt by the parcel in the frame of reference rotating with the earth is

$$\vec{F}_{SB} = \frac{U_{ROT}^2}{r}\hat{n} = \Omega^2 r \hat{n}, \qquad (6.17)$$

where $\hat{n}$ is the unit vector perpendicular to and pointing away from the axis of rotation (Figure 6.3). Gray arrows in Figure 6.3 represent forces in the rotating frame of reference that act on a parcel in solid body rotation. These are gravity, $\vec{F}_{GRAV}$, and the centrifugal force, $\vec{F}_{SB}$. These two forces do not balance, so some other net force is required to keep the parcel over the same location on the earth's surface. Note that solid body rotation is the result of a particular balance of forces and not a case of zero net force acting on a parcel.

In the formulation of the equations of motion for the atmosphere and oceans, $\vec{F}_{SB}$ is absorbed into $\vec{F}_{GRAV}$. *Effective gravity*, $\vec{g}_{EFF}$, is defined as

$$\vec{g}_{EFF} = \vec{F}_{GRAV} + \vec{F}_{SB} \qquad (6.18)$$

and drawn in Figure 6.3. The difference between $\vec{F}_{GRAV}$ and $\vec{g}_{EFF}$ is greatly exaggerated in Figure 6.3, and we will neglect $\vec{F}_{SB}$ compared with $\vec{g}$, so that $\vec{F}_{GRAV} \approx \vec{g}_{EFF}$.

Now, consider a parcel with an eastward zonal velocity, $u > 0$, relative to the earth's surface. In the absolute frame of reference, this parcel is traveling

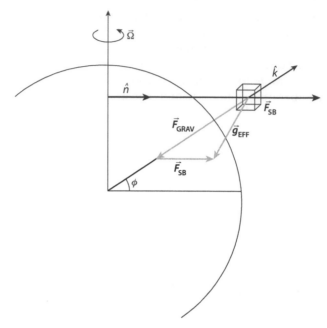

Figure 6.3. Centrifugal acceleration for solid body rotation.

around the same circle as the parcel in solid body rotation, but at a faster rate—it is "super rotating." The centrifugal force acting on this parcel, $\vec{F}_{CENT}$, is greater than that acting on the parcel in solid body rotation, $\vec{F}_{SB}$, and can be written

$$\vec{F}_{CENT} = \vec{F}_{SB} + \vec{F}'_{CENT} = \Omega^2 r\hat{n} + \vec{F}'_{CENT}, \tag{6.19}$$

where $\vec{F}'_{CENT}$ is the difference in centrifugal acceleration that arises when $u \neq 0$ and Eq. 6.17 has been used. If $u > 0$, then $\vec{F}'_{CENT}$ points in the $+\hat{n}$ direction, and if $u < 0$ (the case of "sub-rotation"), $\vec{F}'_{CENT}$ points in the $-\hat{n}$ direction.

By analogy with Eq. 6.17, the full centrifugal acceleration acting on the parcel can also be written

$$\vec{F}_{CENT} = \frac{U_{ABS}^2}{r}\hat{n} = \frac{(U_{ROT} + u)^2}{r}\hat{n} = \Omega^2 r\hat{n} + \left(2\Omega u + \frac{u^2}{r}\right)\hat{n}. \tag{6.20}$$

Comparing Eq. 6.20 with Eq. 6.19, we see that

$$\vec{F}'_{CENT} = \left(2\Omega u + \frac{u^2}{r}\right)\hat{n}. \tag{6.21}$$

$\vec{F}'_{CENT}$ can be decomposed into a meridional component, $F^y_{CENT}\hat{j}$, and a vertical component, $F^z_{CENT}\hat{k}$, as diagrammed in Figure 6.4. In local Cartesian coordinates, the unit vector $\hat{n}$ is

$$\hat{n} = -\sin\phi\hat{j} + \cos\phi\hat{k}. \tag{6.22}$$

With $r = a\cos\phi$, Eq. 6.21 becomes

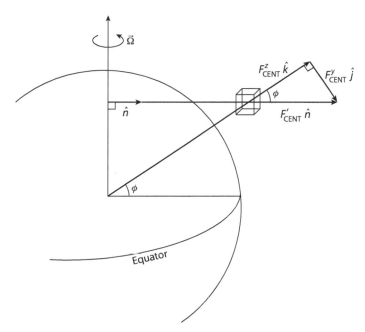

Figure 6.4. Decomposition of the centrifugal acceleration into local Cartesian coordinates.

$$\vec{F}'_{\text{CENT}} = \left(-2\Omega u \sin\phi - \frac{u^2 \tan\phi}{a} \right)\hat{j} + \left(2\Omega u \cos\phi + \frac{u^2}{a} \right)\hat{k}. \qquad (6.23)$$

This acceleration, representing the centrifugal force per unit mass associated with nonzero zonal velocity, is part of the Coriolis force.

A parcel with meridional velocity v relative to the earth's surface will also experience centrifugal acceleration. This acceleration, denoted $\vec{F}''_{\text{CENT}}$, will be oriented in the local vertical, the $\hat{k}$ direction. Because there is no component of the earth's rotation in the meridional direction, this centrifugal acceleration is simply

$$\vec{F}''_{\text{CENT}} = \frac{v^2}{a}\hat{k}. \qquad (6.24)$$

THE FULL CORIOLIS FORCE

The full Coriolis force, $\vec{F}_{\text{COR}}$, is constructed by adding the component that accounts for conservation of absolute angular momentum (Eq. 6.16) to the components that account for centrifugal accelerations due to zonal and meridional motion, Eqs. 6.23 and 6.24, respectively:

$$\vec{F}_{\text{COR}} = \left(2\Omega v \sin\phi + \frac{uv}{a}\tan\phi - 2\Omega w \cos\phi - \frac{uw}{a} \right)\hat{i} + \left(-2\Omega u \sin\phi - \frac{u^2 \tan\phi}{a} \right)\hat{j}$$
$$+ \left(2\Omega u \cos\phi + \frac{u^2 + v^2}{a} \right)\hat{k}. \qquad (6.25)$$

An alternative approach to deriving the full Coriolis force is to perform a formal coordinate translation from the inertial to the rotating frame of reference. The physically motivated derivation provided here gives the exact same results.

An approximate expression for the Coriolis force can be obtained from Eq. 6.25 by neglecting terms inversely proportional to the radius of the earth (a) and terms involving vertical velocity, since $w \ll u$ and v (see chapter 2 and exercise 6.4). The vertical component of the full Coriolis force can also be neglected in this approximation, since it is small compared with gravity (exercise 6.4). These approximations leave two dominant terms in Eq. 6.25, and

$$\vec{F}_{COR} \approx 2\Omega v \sin\phi \hat{i} - 2\Omega u \sin\phi \hat{j} = f(v\hat{i} - u\hat{j}), \tag{6.26}$$

where

$$f \equiv 2\Omega \sin\phi \tag{6.27}$$

is known as the *Coriolis parameter*. In vector form,

$$\vec{F}_{COR} \approx -f\hat{k} \times \vec{v}. \tag{6.28}$$

In the approximation to the full Coriolis force given in Eq. 6.26, the meridional component derives from centrifugal forces and the zonal component from conservation of absolute angular momentum.

Note that the Coriolis force couples the two horizontal directions of motion. According to Eq. 6.26, if a parcel has a non-zero meridional velocity v (in the rotating frame of reference), it will acquire a zonal acceleration due to conservation of absolute angular momentum. If the parcel has a non-zero zonal velocity u, it will be subjected to a meridional acceleration due to centrifugal forces.

6.2 PRESSURE GRADIENT FORCE

Pressure gradient forces are also extremely important in determining winds and ocean currents. These forces arise when pressure—or geopotential height—fields are not uniform. In that case, air and water parcels experience a pressure gradient force directed to lower values, or down the gradient.

Consider a volume V of air or water with pressure p_1 acting on the left face located at n_1, and pressure p_2 acting on the right face located at n_2, as depicted in Figure 6.5. The area of each face of the volume is A, and the volume contains a mass of material m with density ρ. The direction indicated by unit vector $\hat{n}$ can be any direction, horizontal or vertical or some combination.

The pressure on the left side of the volume, p_1, is the force per unit area exerted by the surrounding air or water, and p_2 is the pressure on the right face. If $p_1 = p_2$, the parcel will not move (accelerate) in the $\hat{n}$ direction. If $p_1 > p_2$, the parcel will accelerate to the right ($+\hat{n}$ direction), and if $p_1 < p_2$, the parcel will accelerate to the left ($-\hat{n}$ direction). The magnitude of the acceleration depends on the difference in p_1 and p_2, and the direction is determined by which is bigger.

To translate this concept into a mathematical expression for use in Eqs. 6.3– 6.5, we define $\vec{F}_{PGF}$ as the force per unit mass acting on the parcel due to pressure gradients. Then,

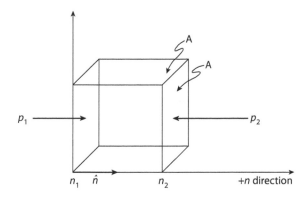

Figure 6.5. The pressure gradient force.

$$\vec{F}_{PGF} = \frac{(\text{force acting at } n_1) - (\text{force acting at } n_2)}{\text{mass of parcel}} \hat{n}. \qquad (6.29)$$

But since pressure is force per unit area,

$$\text{force acting at } n_1 = Ap_1,$$

and likewise for the right face at n_2. Therefore,

$$\vec{F}_{PGF} = \frac{A}{m}(p_1 - p_2)\hat{n} = -\frac{1}{\rho}\frac{(p_2 - p_1)}{(n_2 - n_1)}\hat{n}, \qquad (6.30)$$

since

$$\rho = \frac{m}{V} = \frac{m}{A\Delta n} = \frac{m}{A(n_1 - n_2)} \Rightarrow \frac{A}{m} = \frac{1}{\rho(n_1 - n_2)}. \qquad (6.31)$$

Taking the differential limit of Eq. (6.30), we obtain the final expression for the pressure gradient force:

$$\vec{F}_{PGF} = -\frac{1}{\rho}\frac{\partial p}{\partial n}\hat{n}. \qquad (6.32)$$

Note the negative sign in Eq. 6.32, which indicates that the direction of the pressure gradient force is down the gradient, that is, from high to low values. In local Cartesian coordinates, the *horizontal pressure gradient force* is

$$\vec{F}_{PGF}^{x,y} = -\frac{1}{\rho}\frac{\partial p}{\partial x}\hat{i} - \frac{1}{\rho}\frac{\partial p}{\partial y}\hat{j} \qquad (6.33)$$

and the vertical pressure gradient force is

$$\vec{F}_{PGF}^{z} = -\frac{1}{\rho}\frac{\partial p}{\partial z}\hat{k}. \qquad (6.34)$$

6.3 HYDROSTATIC BALANCE

Because pressure decreases with height in the earth's atmosphere and oceans, $\partial p/\partial z < 0$ and the vertical pressure gradient force (Eq. 6.34) is directed upward (positive $+\hat{k}$ direction). So why doesn't all the air leave the planet? The answer is that gravity exerts a compensating downward force. *Hydrostatic balance* occurs when the downward force of gravity balances the upward force of the vertical pressure gradient:

$$-\frac{1}{\rho}\frac{\partial p}{\partial z} = g \Rightarrow \frac{\partial p}{\partial z} = -\rho g. \tag{6.35}$$

Equation 6.35 is an excellent approximation for the vertical balance of forces in both the atmosphere and the oceans on large scales of motion, away from the surface where frictional accelerations are strong.

Combining the hydrostatic and adiabatic conditions (section 5.2) for an ideal gas allows us to calculate the atmospheric *adiabatic lapse rate*. Taking the natural log (ln) of Eq. 5.6 and the Lagrangian derivative of the resulting equation, we obtain

$$\frac{1}{\Theta}\frac{d\Theta}{dt} = \frac{1}{T}\frac{dT}{dt} - \frac{R}{c_p p}\frac{dp}{dt} = 0 \quad \Rightarrow \quad \frac{1}{T}\frac{dT}{dt} = \frac{R}{c_p p}\frac{dp}{dt}. \tag{6.36}$$

To focus on the height dependence, we consider $T = T(z)$ and $p = p(z)$ alone (T and p depend only on elevation). Then, $d/dt = w(\partial/\partial z)$ (see Appendix B) and Eq. 6.36 becomes

$$\Gamma_{\text{ADIABATIC}} = -\frac{\partial T}{\partial z} = -\frac{RT}{c_p p}\frac{\partial p}{\partial z} = \frac{g}{c_p} = 9.8 \text{ K/km} \tag{6.37}$$

for an ideal gas under adiabatic, hydrostatic conditions. In other words, as a parcel of air rises adiabatically in a hydrostatic atmosphere, it will cool 9.8 K for every 1 km increase in elevation, and we would observe a lapse rate of 9.8 K/km. Figure 2.8 shows that the observed lapse rate in the troposphere is roughly half the adiabatic lapse rate, which suggests that diabatic heating (such as the radiative heating processes and the turbulent heat fluxes discussed in chapter 5) is important.

The hydrostatic relation (Eq. 6.35) can be integrated vertically to derive an equation for p as a function of z in the atmosphere. A very simple analytical form of this relationship can easily be derived if two assumptions are made. The first is that the atmosphere behaves as an ideal gas (section 5.1). This is an excellent assumption. The second assumption is that the atmosphere is isothermal. This is a terrible assumption, as Figures 2.8 and 2.9 demonstrate, but it is useful for a first-order derivation of how pressure depends on elevation in the atmosphere.

Using the ideal gas law (Eq. 5.1) to eliminate density from the hydrostatic relation (Eq. 6.35) and rearranging, we have

$$\frac{dp}{p} = -\frac{g}{RT}dz. \tag{6.38}$$

Integrating from the surface, where $z = 0$ and $p = p_0$, to some level (z, p) in an isothermal atmosphere we get

$$p = p_0 e^{-z/H}, \tag{6.39}$$

where $H \equiv RT/g$ is the *atmospheric scale height*. When $z = H$, $p = p_0/e = p_0/2.718$, so H is an *e*-folding distance, or the vertical distance over which pressure decreases by a factor of about 3, in an hydrostatic, isothermal atmosphere.

Figure 2.2 shows how pressure and height are related in the earth's atmosphere; this is essentially a graph of Eq. 6.39, with pressure falling off exponentially with height. Note that it is a one-to-one relationship, making pressure a viable choice as a vertical coordinate in place of height. From Figure 2.2 we can estimate that $H \approx 9$ km, since pressure decreases to about one-third of its surface value over about 9 km. The scale height is often used to characterize atmospheric depth because it is not possible exactly to define the top of an atmosphere.

In the ocean, where the ideal gas law cannot be applied, the relationship between pressure and elevation (depth) is different. For the ocean we can make the incompressibility assumption that density is constant with depth; this is an accurate assumption below the pycnocline, that is, below about 1000 km depth. Integrating the hydrostatic balance equation (Eq. 6.35) from the surface of the ocean, where $p = p_0$ and $z = 0$, to some depth z, where $z < 0$, we obtain

$$\int_{p_0}^{p} dp = -\int_{0}^{z} \rho g \, dz = -\rho g z \Rightarrow p = p_0 - \rho g z. \tag{6.40}$$

So, in the hydrostatic, incompressible ocean, pressure is a linear function of depth.

We now turn to the horizontal directions. When pressure is used as a vertical coordinate, the horizontal pressure gradient force can be expressed as a function of geopotential (or geopotential height). According to Eq. 2.3, $g \, dz = d\Phi$, so the hydrostatic relation (Eq. 6.35) can be written as

$$g|\Delta z| = |\Delta\Phi| = \frac{|\Delta p|}{\rho}. \tag{6.41}$$

Then, from Eq. 6.33,

$$\vec{F}_{PGF} = -\frac{\partial\Phi}{\partial x}\hat{i} - \frac{\partial\Phi}{\partial y}\hat{j}, \tag{6.42}$$

where the horizontal derivatives must be taken on a pressure surface (holding p constant) instead of a z surface (constant elevation). Recall that regions of low geopotential height are regions in which the height of a pressure level is low, that is, where isobaric surfaces dip closer to the surface. Because pressure decreases with elevation in the atmosphere, regions with low geopotential height are also regions of low pressure.

The hydrostatic relation and the idea gas law can be used to express the relationship between geopotential height and temperature in the atmosphere. Using the ideal gas law in Eq. 6.38, we obtain

$$d\Phi = -\frac{RT}{p} dp. \tag{6.43}$$

Integration over a pressure interval from p_1 to p_2 gives

$$\Phi_1 - \Phi_2 = R \int_{p_2}^{p_1} T \, d\ln p \quad \Rightarrow \quad Z_1 - Z_2 = \frac{R}{g} \int_{p_2}^{p_1} T \, d\ln p. \qquad (6.44)$$

Equation 6.44 indicates that the distance between two geopotential height levels (known as the *thickness*) is proportional to the mass-weighted average temperature between the two layers.

6.4 GEOSTROPHIC BALANCE

For large spatial scales (hundreds of kilometers or more), away from the surface and the equator, the largest horizontal forces acting on parcels in the atmosphere are horizontal pressure gradient forces and Coriolis forces. This is also the case in the ocean for space scales greater than about 50 km, away from the coasts and below the mixed layer. When the horizontal pressure gradient force is balanced by the Coriolis force, the flow is said to be in *geostrophic balance*.

The wind or ocean current speed for the case of geostrophic balance is known as the *geostrophic velocity* and denoted $\vec{v}_G = u_G \hat{i} + v_G \hat{j}$. It can be calculated by setting the magnitude of the horizontal pressure gradient force (Eq. 6.33) equal to that of the (approximate) Coriolis force (Eq. 6.26). In the x-direction,

$$2\Omega v \sin\phi - \frac{1}{\rho} \frac{\partial p}{\partial x} = 0 \quad \Rightarrow \quad v_G = \frac{1}{\rho f} \frac{\partial p}{\partial x}, \qquad (6.45)$$

and in the y-direction,

$$-2\Omega u \sin\phi - \frac{1}{\rho} \frac{\partial p}{\partial y} = 0 \quad \Rightarrow \quad u_G = -\frac{1}{\rho f} \frac{\partial p}{\partial y}. \qquad (6.46)$$

In vector form,

$$\vec{v}_G = \hat{k} \times \frac{\nabla p}{\rho f}. \qquad (6.47)$$

Note again that the Coriolis force couples the east/west and north/south directions of motion in the rotating frame of reference. According to Eq. 6.45, meridional geostrophic velocity is generated by *zonal* pressure gradients, and zonal geostrophic velocity is generated by *meridional* pressure gradients according to Eq. 6.46.

With pressure as a vertical coordinate,

$$\vec{v}_G = u_G \hat{i} + v_G \hat{j} = -\frac{1}{f} \frac{\partial \Phi}{\partial y} \hat{i} + \frac{1}{f} \frac{\partial \Phi}{\partial x} \hat{j}, \qquad (6.48)$$

or

$$f\vec{v}_G = \hat{k} \times \nabla_p \Phi. \qquad (6.49)$$

The magnitude of the geostrophic wind velocity, V_G, in p-coordinates is

$$V_G = \sqrt{u_G^2 + v_G^2} = \left| \frac{1}{f} \frac{\partial \Phi}{\partial n} \right|. \qquad (6.50)$$

According to Eq. 6.50, higher geostrophic wind speeds occur where horizontal geopotential height gradients are large. Figure 2.14 portrays a clear example of this relationship since the 200 hPa wind vectors are larger in regions where the geopotential height lines pinch closer together.

According to Eq. 6.49, the geostrophic wind is perpendicular to geopotential (and geopotential height) gradients in the horizontal plane and, therefore, flows parallel to geopotential height lines. This explains the relationship between the wind direction and geopotential height lines in Figures 2.13 and 2.14. The relationship breaks down at low latitudes because the Coriolis parameter, f, becomes small and the Coriolis force is less able to balance pressure gradients forces. At these low latitudes, the flow is more directly down the pressure gradient.

As diagrammed in Figure 6.6, application of the right-hand rule to Eq. 6.49 in the Northern Hemisphere ($f > 0$) indicates that low heights (or pressures) are located to the left of the wind direction. In the Southern Hemisphere, where $f < 0$, low heights lie to the right of the geostrophic wind vector. Again, these relationships are seen clearly in Figures 2.13 and 2.14 and explain why cyclonic flow (flow about a low-pressure system), for example, is counterclockwise in the Northern Hemisphere but clockwise in the Southern Hemisphere.

One great advantage of using pressure instead of elevation as a vertical coordinate is that density does not appear explicitly in the geostrophic wind relation. This makes it easy to derive an expression for the geostrophic wind shear (the vertical change in the geostrophic wind) by taking derivatives with respect to pressure. From Eq. 6.48,

$$\frac{\partial u_G}{\partial p} = -\frac{1}{f}\frac{\partial}{\partial p}\frac{\partial \Phi}{\partial y} = \frac{1}{f}\frac{\partial}{\partial y}\left(\frac{RT}{p}\right) \Rightarrow \frac{\partial u_G}{\partial \ln p} = \frac{R}{f}\frac{\partial T}{\partial y}. \tag{6.51}$$

Similarly,

$$\frac{\partial v_G}{\partial \ln p} = -\frac{R}{f}\frac{\partial T}{\partial x}. \tag{6.52}$$

In vector form,

$$\frac{\partial \vec{v}_G}{\partial \ln p} = -\frac{R}{f}\hat{k} \times \nabla_p T. \tag{6.53}$$

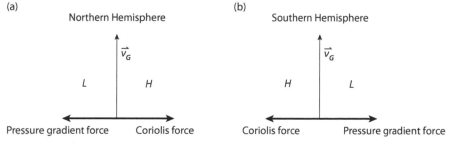

Figure 6.6. Balance of forces in the Northern Hemisphere and the Southern Hemisphere for geostrophic conditions.

Equations 6.51 and 6.52 are known as the *thermal wind equations*, which is confusing because they are expressions for the wind shear and not the wind itself. Equation 6.51 indicates that a negative meridional temperature gradient, $\partial T/\partial y < 0$, such as is found on average in Northern Hemisphere middle latitudes where temperature decreases with increasing latitude, is associated with winds that become increasingly westerly with elevation, that is, $\partial u_G/\partial z > 0$ or $\partial u_G/\partial \ln p < 0$. In the Southern Hemisphere, with $f < 0$ and $\partial T/\partial y > 0$, winds also become more westerly with elevation. This relationship is borne out in observations (see Figs. 2.9 and 2.10). The geostrophic relationship between the wind velocity and meridional temperature gradients explains the location of the westerly jets and their seasonal variation.

6.5 FRICTION

Frictional forces are important close to the surface and, in the oceans, close to the coasts where they act opposite to the direction of the flow. To understand one role of friction in atmospheric flows, consider a parcel of air in geostrophic balance near a low in the Northern Hemisphere. Adding friction to the balance of forces decreases the wind speed (Figure 6.7) and, as a consequence, reduces the magnitude of the Coriolis force since it is proportional to the velocity (Eq. 6.28). The result is that the pressure gradient force is stronger than the Coriolis force, and the flow accelerates down the pressure gradient. For this reason, the low-level flow around a low-pressure system near the surface has a component directed into the low. The air converges, and conservation of mass requires upward motion. If the air is moist, this movement can lead to clouds and precipitation. Similarly, it is common to find divergence in the vicinity of a high due to the effects of friction, and the sinking air tends to be clear and cool.

Frictional acceleration is also important in the ocean mixed layer, where the effects of surface winds are strong. The horizontal equations of motion for this case are derived in chapter 8.

On or very close to the equator, the Coriolis force tends to zero and geostrophic flow is no longer sustained. Here, the main balance of forces is between

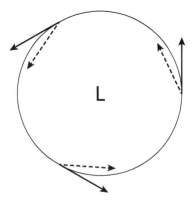

Figure 6.7 Low-level flow around a low in the Northern Hemisphere to illustrate frictional convergence. Solid arrows indicate the direction of the geostrophic flow; dashed arrows indicate the direction of the flow when frictional acceleration is added to the geostrophic balance.

friction and pressure gradient forces. The flow is generally directed perpendicular to the isobars or isoheights, down the pressure gradients into lows or away from highs. More mass flows into regions of low pressure than at higher latitudes, and lows are substantially weaker.

6.6 THE MOMENTUM EQUATIONS

The expressions developed for the Coriolis and pressure gradient forces can be used to write the horizontal equations of motion (Eqs. 6.3 and 6.4) in z-coordinates as

$$\frac{du}{dt} = fv - \frac{1}{\rho}\frac{\partial p}{\partial x} + F_F^x \tag{6.54}$$

and

$$\frac{dv}{dt} = -fu - \frac{1}{\rho}\frac{\partial p}{\partial y} + F_F^y \tag{6.55}$$

where the approximate form of the Coriolis force from Eq. 6.26 is used. If we use the Eulerian perspective in local Cartesian coordinates (Appendix C), the local wind or current velocity is accelerated according to

$$\frac{\partial u}{\partial t} = -\vec{v}\cdot\nabla u + fv - \frac{1}{\rho}\frac{\partial p}{\partial x} + F_F^x \tag{6.56}$$

and

$$\frac{\partial v}{\partial t} = -\vec{v}\cdot\nabla v - fu - \frac{1}{\rho}\frac{\partial p}{\partial y} + F_F^y. \tag{6.57}$$

When the local time derivatives of u and v, the advection terms, and friction are small, the flow is geostrophic. This highlights the fact that that geostrophic wind is steady, or nonaccelerating.

The vertical equation of motion is

$$\frac{\partial w}{\partial t} = -\vec{v}\cdot\nabla w - g - \frac{1}{\rho}\frac{\partial p}{\partial z} + F_F^z. \tag{6.58}$$

Hydrostatic balance (Eq. 6.35) occurs when the local time rate of change of w, advection of vertical velocity, and friction are negligible.

6.7 EXERCISES

6.1. Calculate the pressure gradient force for the system drawn in Figure 6.8. Write vector equations expressing your answers using the local Cartesian coordinate system, and also draw a vector indicating the direction of the pressure gradient force. Assume that density is 1.0 kg/m³.

6.2. Calculate the zonal velocity relative to the rotating earth, u, attained by a parcel of air in moving from the equator to 30°N latitude conserving

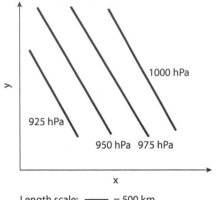

Figure 6.8 Isobars in the horizontal plane.

absolute angular momentum. Compare this velocity with observed zonal velocities in chapter 2.

6.3. The equation $dM/dt = 0$ expresses the principle of conservation of absolute angular momentum. Substituting Eq. 6.13 into this equation, carry through the differentiation to find the zonal component of the Coriolis force, Eq. 6.16. (*Hint*: Do not make the thin atmosphere approximation until after you have carried through the differentiation and set $dz/dt = w$.)

6.4. Estimate the magnitude of each term in Eq. 6.25 using the figures of chapter 2 to choose reasonable orders of magnitude for values for the dependent variables. (For example, set $u \sim v \sim 10$ m/s.) To focus on large space scales, let $dx \sim dy \sim 1000$ km and $dz \sim 10$ km, which is about the full depth of the troposphere. Make one calculation for middle latitudes ($\phi \sim 45°$N or $45°$S) and observe whether the first-order balance of forces changes in the tropics ($\phi \sim 10°$N or $10°$S).

6.5. The peak elevation for Mount Rainier is 14,410 feet. Assuming that the atmosphere is hydrostatic and isothermal at 270 K, estimate the surface pressure on the top of Mount Rainier. Be sure to check that your answer is reasonable.

6.6. To what degree of accuracy can the vertical component of the Coriolis force (per unit mass) be neglected compared with the acceleration due to gravity?

6.7. Across Texas, the meridional wind is steady and uniform at $+3$ m/s (a southerly wind). The zonal wind at Austin is measured at 4 m/s and at San Antonio, about 130 km to the south, the zonal wind is 6 m/s. At what rate will the zonal wind speed at Austin be accelerated due to advection?

6.8. Refer to the figures of chapter 2 to find a reasonable value for the vertically averaged tropospheric meridional temperature gradient in middle latitudes (say, $40°$S–$60°$S) during the Southern Hemisphere winter. Using this value, integrate the zonal thermal wind equation (Eq. 6.51) from the surface to 200 hPa to estimate the zonal wind speed at 200 hPa. Assume that the surface wind speed is 6 m/s. Compare your calculated zonal wind speed with the observed. Are they similar? What might be sources of difference?

6.9. Figure 6.9 is an idealization of the geopotential height distribution at 200 hPa over eastern Asia between $30°$N and $60°$N in July (see Fig. 2.3b).

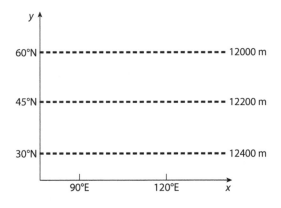

Figure 6.9 Geopotential height contours in the horizontal plane.

Derive an expression for the geostrophic wind, expressing your answer in vector form. Draw the geostrophic wind vector on a sketch of the geopotential height distribution, including an indication of the scale, and compare your answer with Fig. 2.14. To what accuracy is the observed wind approximated by the geostrophic wind?

7
ATMOSPHERIC CIRCULATIONS

There are fundamental differences between atmospheric circulations in middle and high latitudes and in the tropics because of the latitude dependence of the Coriolis force (Exercise 6.4). Poleward of about 20°latitude, depending on the season, the large-scale flow is approximately in geostrophic balance, as seen in the figures of chapter 2. At low latitudes, thermally driven circulation systems dominate in the presence of weaker Coriolis forces and strong condensational heating.

7.1 THERMALLY DIRECT CIRCULATIONS

Thermally direct circulations transfer heat from warmer regions to cooler regions. They arise in association with differences in surface temperature and, therefore, low-level geopotential height and pressure gradients.

Figure 7.1 illustrates the basic features of a thermally direct circulation. The dashed lines indicate geopotential height contours and the arrows show the direction of the flow. Four regions are numbered to facilitate discussion.

The atmospheric temperature over the warmer surface (Region 1) is increased by enhanced heat fluxes from the surface (chapter 5) and, as a result, the thickness (distance between geopotential height lines; see Eq. 6.44) is greater than the thickness over the cooler surface (Region 3). The resulting horizontal geopotential height (and pressure) gradients drive flow down the gradients in Regions 2 and 4. The circuit is closed as air subsides over the surface high and rises over the surface low.

Part of the heat transport by a thermally direct circulation takes place in the form of sensible heat fluxes. Horizontal sensible heat fluxes occur when warm air is transported to a cooler region, and when cooler air is transported to a warmer region. As illustrated in Figure 7.1, the horizontal branches (Regions 2 and 4) of the circulation are responsible for these sensible heat transports.

Latent heat fluxes also play an important role in thermally direct circulation systems. Strong evaporation occurs in Region 4. The air is dry, which enhances evaporation (Eq. 5.21), and the release of latent heat as a result of convection is suppressed by *subsidence* (sinking air) in the down-branch of the circulation. Thus, the specific humidity of the air increases as it flows into the thermal low, where it rises to form the up-branch of the thermally direct circulation. As the warmed parcels of moist air rise they adjust to the exponentially decreasing environmental pressure (Eq. 6.39) and expand and cool adiabatically (Eq. 6.37). When the temperature decrease is sufficiently large, the parcels reach saturation specific humidity (Fig. 2.32) and the water vapor in the parcels condenses. The

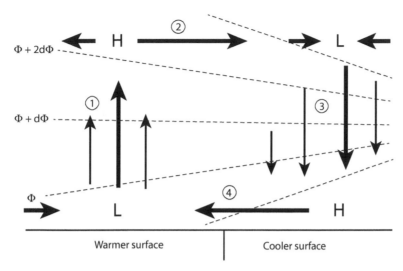

Figure 7.1 A thermally direct circulation system.

resulting release of latent heat into the middle troposphere further increases the air temperature, reducing density (Eq. 5.1) and adding buoyancy, so the parcels continue to rise. In this way, the latent heat flux transports heat vertically. With respect to latitude, however, the latent heat flux works "backward," with cooling by evaporation over the cooler surface (Region 4) and the release of latent heat over the warmer surface (Region 1).

Note how the sensible and latent heat fluxes work together. The release of latent heat in the midtroposphere converts latent to sensible heat, feeding into the horizontal upper-tropospheric sensible heat flux. The overall result is the net transport of heat to cooler regions and higher elevations.

THE HADLEY CIRCULATION

A thermally direct circulation known as the *Hadley circulation* covers the full depth of the troposphere and approximately half the earth's surface area, between 30°N and 30°S. It is a large-scale, overturning circulation (Fig. 7.1) that exists only in the zonal mean, as discussed below. The surface temperature gradient that drives the Hadley circulation is meridional, so the circulation is fundamentally due to the latitudinal dependence of the solar heating and, ultimately, the (approximately) spherical shape of the planet. Hadley circulations occur on other planets as well, most notably Mars, Jupiter, and Venus.

Climatological meridional (dark contours) and vertical p-velocity (light contours) winds are together in Figure 7.2. Both variables are averaged over the full year and across longitudes to form the zonal mean. Arrows indicate the direction of the flow, and regions of high and low geopotential heights (high and low relative to the same level) are indicated. Note the similarity to Figure 7.1. The warmer surface near the equator is co-located with the region of rising

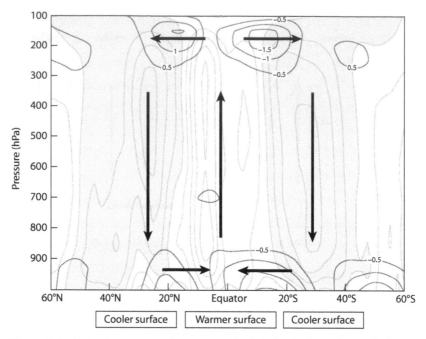

Figure 7.2. Annual mean, zonal mean meridional wind velocity (m/s; dark gray contours) and vertical p-velocity (hPa/s; light gray contours). Shading indicates downward motion. The contour interval for vertical velocity is 5×10^{-3} Pa/s. This plot is a superposition of Figs. 2.11 and 2.12.

motion, and subsidence occurs over the cooler surfaces at subtropical latitudes in both hemispheres.

Figure 7.2 provides a rudimentary view of the Hadley circulation, pieced together by examining the zonal mean meridional and vertical velocity fields jointly. A better way to envision and quantify the Hadley circulation is to define a stream function. Consider the three-dimensional velocity field in local Cartesian p-coordinates, averaged around longitude,

$$[\vec{v}] = [u]\hat{i} + [v]\hat{j} + [\omega]\hat{k}, \tag{7.1}$$

where the square brackets denote the zonal average. Because longitudinal (x-coordinate) dependence is removed by taking the zonal mean, $\partial[u]/\partial x = 0$ and the divergence of the zonal mean flow is

$$\nabla \cdot [\vec{v}] = \frac{\partial[v]}{\partial y} + \frac{\partial[\omega]}{\partial p} = 0. \tag{7.2}$$

According to Eq. 7.2, convergence in the meridional direction must be balanced by vertical divergence. In other words, only one independent variable, either $[v]$ or $[\omega]$, is needed to define the zonally averaged flow field. With a little mathematical manipulation, a stream function, Ψ, can be used as that one variable.

A simple way to define the stream function is with the following pair of equations:

$$[v] = \frac{\partial \Psi}{\partial p} \tag{7.3}$$

and

$$[\omega] = -\frac{\partial \Psi}{\partial y}, \tag{7.4}$$

which have been constructed so that Eq. 7.2 is satisfied. A variation on this idea is commonly used to quantify the Hadley circulation. Spherical coordinates (Appendix B) are used because the space scale of the circulation system is large, and the constant g is added so the units of Ψ are kg/s, that is, a mass flux. The *Stokes stream function* is defined by

$$[v] = \frac{g}{2\pi a \cos\phi} \frac{\partial \Psi}{\partial p} \tag{7.5}$$

and

$$[\omega] = -\frac{g}{2\pi a^2 \cos\phi} \frac{\partial \Psi}{\partial \phi}, \tag{7.6}$$

which satisfy the divergence equation

$$\nabla \cdot \vec{v} = \frac{1}{a \cos\phi} \frac{\partial (\cos\phi [v])}{\partial \phi} + \frac{\partial [\omega]}{\partial p} = 0. \tag{7.7}$$

Because vertical velocities are small in the atmosphere and, therefore, difficult to observe accurately, the Stokes stream function is calculated from observations of zonally averaged meridional velocity with the assumption (boundary condition) that the stream function is zero at the tropopause. Integrating Eq. 7.5 from the tropopause *down* to some level p at a fixed latitude gives

$$\frac{2\pi a \cos\phi}{g} \int_{p_{\text{TROP}}}^{p} [v(\phi,p)]dp = \int_{0}^{\Psi(\phi,p)} d\psi \;\Rightarrow\; \Psi(\phi,p) = \frac{2\pi a \cos\phi}{g} \int_{p_{\text{TROP}}}^{p} [v(\phi,p)]dp. \tag{7.8}$$

The integral on the right-hand side of Eq. 7.8 can be calculated numerically from an observed meridional velocity field. $\Psi(\phi,p)$ is a mass transport (kg/s), equal to the rate at which air is transported meridionally in the region between pressure level p and the tropopause by the zonal mean circulation.

Figure 7.3 shows observed monthly mean values of the Stokes stream function. During December, January, February, and March, the mean meridional circulation consists of a single Hadley cell centered near 15°N with its up-branch in the Southern Hemisphere tropics. During May, June, July, August, and September the pattern is reversed, and the Hadley cell is centered in the Southern Hemisphere (the winter hemisphere) with its up-branch in the Northern Hemisphere (the summer hemisphere). The pattern is symmetric about the equator—with two Hadley cells of approximately equal strength—only during three months (April, October, and November) and in the annual mean (Fig. 7.2).

The following features of the climate system are directly related to the Hadley circulation:

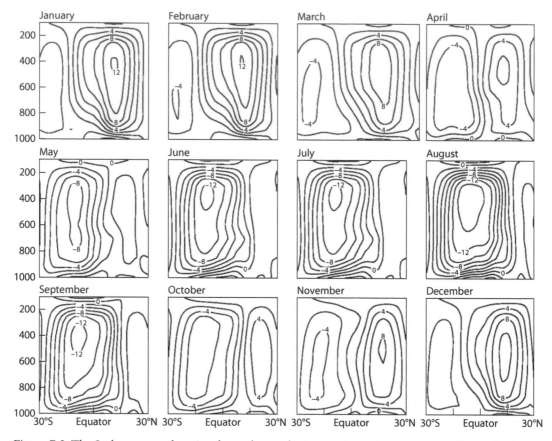

Figure 7.3. The Stokes stream function for each month. Positive (negative) contours indicate clockwise (counterclockwise) circulation. Contour intervals are 2×10^{10} kg/s and vertical axes are in hPa.

- Rising air in the up-branch of the Hadley circulation near the equator supports the zonal mean precipitation maximum, the ITCZ (Figs. 2.25 and 2.26), which is located in the summer hemisphere (in the zonal mean). Air rises in concentrated areas of convection and sinks weakly between convective towers.
- Precipitation minima (deserts) are found in the subtropics (Figs. 2.25 and 2.26), where convection is suppressed in the broad subsiding down-branches of the Hadley circulation.
- The location of the zonal mean evaporation maxima in the subtropics (Figs. 2.28 and 2.29) is the result of the low-level equatorward "return flow" of the Hadley circulation, which consists of dry air moving equatorward in an environment of large-scale subsidence and, therefore, suppressed convection. This connection to the Hadley circulation explains why the zonal mean precipitation and evaporation maxima are not located at the same longitude.
- The salinity maximum in the subtropics (Figs. 2.19 and 2.20) is also related to the Hadley circulation; it is a consequence of excess evaporation over precipitation.

- Easterly low-level flow (Fig. 2.13) occurs in the tropics and subtropics (the trade wind regime) as Coriolis accelerations act on the low-level equatorward return flow of the Hadley circulation.

MONSOON CIRCULATIONS

A monsoon climate is technically defined as a 180° change in the direction of the low-level wind from summer to winter. The winter monsoon features dry conditions and flow off the land surface, and the summer monsoon is the rainy period with moist, onshore flow. However, this definition of a monsoon climate is not always held strictly, and many tropical climates are defined as "monsoon climates" when they exhibit a strong summer rainy season. Monsoon regions include Southeast Asia and India, the southwestern United States, West Africa, East Africa, Australia, and South America.

Both winter and summer monsoon circulations are thermally direct, with the surface temperature contrast provided by continentality. Thus, the fundamental cause of monsoon climates is the different heat capacities of the land and ocean, which causes the oceans to be warmer than the land in the winter months and cooler in the summer months (Fig. 2.7).

Figure 7.4a shows the low-level (900 hPa) flow and geopotential heights across northern Africa, the Arabian Sea, and India during January, the time of the winter monsoon. Relatively high geopotential heights cover subtropical latitudes, with lower geopotential heights to the south. Strong winds are directed off the Indian subcontinent into the Arabian Sea and the Bay of Bengal—this is the winter monsoon flow. Over West Africa, the flow is from the northeast, forming the dry and dusty *Harmattan* of that continent's winter monsoon.

In July (Fig. 7.4b), the low-level wind directions are approximately reversed. Southwesterly flow is in place across West Africa, with southerly flow crossing the Guinean coast (about 5°N). A thermal low, indicated by low 900 hPa geopotential heights, is centered near 20°N. This is the West African summer monsoon system. Over East Africa and the Arabian Sea, the *Somali jet* carries moisture from the Southern Hemisphere across the Horn of Africa and eastward onto the Indian subcontinent to fuel, in part, the Asian monsoon. Rather than being supported by mean flow onto the continent from the Bay of Bengal, the monsoon rains over Bangladesh arrive in the form of storms known as *monsoon depressions*.

The vertical structure of monsoon climates resembles the sketch in Figure 7.1. Near the tropopause in July (Fig. 7.5a), high geopotential heights form over the low geopotential heights seen at 900 hPa (Fig. 7.4b). Southward outflow from the high is diverted to the west by Coriolis forces, and the resulting easterly flow forms the *tropical easterly jet* between 10 and 15°N (Fig. 2.14).

Over West Africa, the thermal low (see Fig. 7.4b) occurs within much drier conditions (Fig. 2.31), so the up-branch of the thermally direct circulation in this region is not enhanced by the release of latent heat as air parcels rise. As a consequence, the system is shallower than over India and the upper-level high occurs at a lower level. This upper-level high is known as the *Saharan high*,

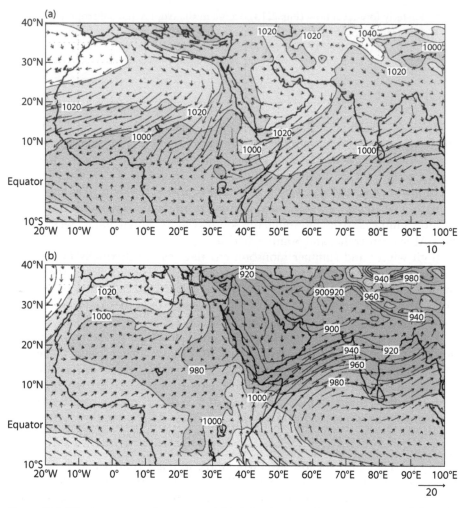

Figure 7.4. Wind vectors and geopotential height contours for (a) January and (b) July at 900 hPa in the vicinity of the African and Indian monsoon systems. Values are extrapolated where the 900 hPa surface is below the physical surface.

and it is centered at about 600 hPa (see Fig. 7.5b). Easterly flow south of the Saharan high is known as the *African easterly jet*.

WALKER CIRCULATIONS

A third example of thermally direct circulations in the atmosphere is the *Walker circulation*. This is a series of east–west overturning cells close to the equator where Coriolis forces are small. Rising branches are located over relatively warm areas (the western warm pool in the Pacific, equatorial South America, and equatorial Africa) and sinking branches are over cooler areas (e.g., the eastern Pacific and Atlantic cold tongues; see Fig. 2.6).

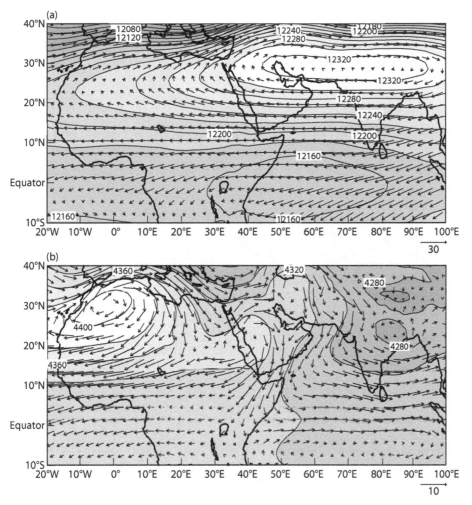

Figure 7.5. Climatology of (a) 200 hPa and (b) 600 hPa geopotential heights and winds in July. Contour intervals are 20 m.

The Walker circulation is particularly well developed in the equatorial Pacific, with rising air in the west and subsidence in the east. Figure 7.6a shows the Pacific vertical p-velocity climatology at 500 hPa, the level at which vertical velocities are typically largest. The down-branch of the Pacific Walker circulation, with positive ω values indicating sinking motion, is located the eastern Pacific and the rising branch is in the west.

The SOI, that is, the strong correlation between surface pressures at Darwin, Australia, and Tahiti (chapter 3), comes about as the Pacific Walker circulation shifts with ENSO. The vertical velocity distribution typical of warm events (El Niño) is displayed in Figure 7.6b. Sinking in the eastern Pacific is weakened and confined south of the equator, and the border between positive and negative ω values is located farther east compared with the climatology. The region

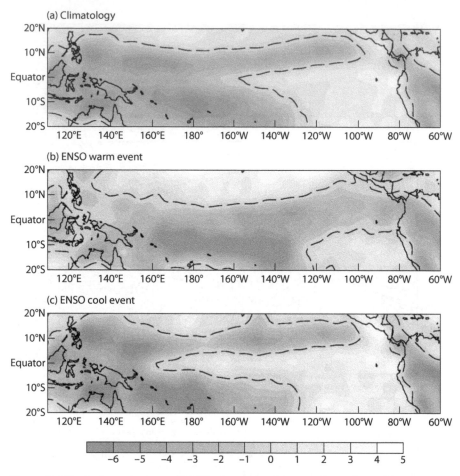

Figure 7.6 Vertical *p*-velocity at 500 hPa from (a) a 50-year climatology, (b) an average for DFJ of the 1982/1983 and 1997/1998 ENSO warm events, and (c) an average for DFJ of the 1975/1976 and 1988/1989 ENSO cool events. Units are 0.01 Pa/s and the 0 line is dashed.

of maximum negative values (upward motion) is located in the central Pacific instead of in the west, and vertical velocities in the western Pacific are weak.

During ENSO cool events (La Niña), sinking motion over the eastern Pacific is enhanced and extends to the west along the equator (Fig. 7.6c). Upward motion in the far western Pacific (120°E–160°E) and over South America is stronger than in the climatology.

The Walker circulation across the equatorial Pacific reinforces the easterly flow near the surface (Fig. 2.13). These low-level easterly winds push warm surface water to the western Pacific, raising sea level there about 20 cm above that in the eastern Pacific. This easterly flow helps maintain the longitudinal sea surface temperature gradient across the Pacific (Fig. 2.15) which, in turn, maintains the Walker circulation. In this way, the atmosphere and ocean are "coupled."

7.2 MIDLATITUDE CIRCULATION SYSTEMS

In middle latitudes, a weak meridional circulation known as the *Ferrel cell* lies between a thermally direct polar cell and the Hadley cell. In the Ferrel cell, warmer air closer to the equator (coincident with the down-branches of the Hadley circulation) sinks, and air at cooler, high latitudes (near 60°N and 60°S, coincident with the up-branches of a polar cell) rises. This means that in middle latitudes, the zonal mean meridional circulation transports heat downward and equatorward—it is *thermally indirect*. Poleward energy transport in middle latitudes is accomplished not by the mean meridional circulation but by the deviations from the zonal mean, or "eddies," in the form of storms and other disturbances.

The *index cycle* is a way of conceptualizing the poleward transport of heat by eddies. Any variable can decomposed into the sum of a zonal mean and a deviation from the zonal mean according to

$$u = [u] + u^* \quad \text{and} \quad v = [v] + v^*, \tag{7.9}$$

where the square brackets denote the zonal mean and the asterisks indicate the deviation from the zonal mean, or "eddy." See Figure 7.7a to visualize these two components of the flow. As illustrated in the figures of chapter 2, $[v]$ is much smaller than $[u]$ in midlatitudes and so is neglected here, but u^* and v^* are comparable.

The index cycle begins with a steady, uniform westerly wind (see Figure 7.7b), that is, $[u]$ is large and u^* and v^* are essentially zero, with cold polar

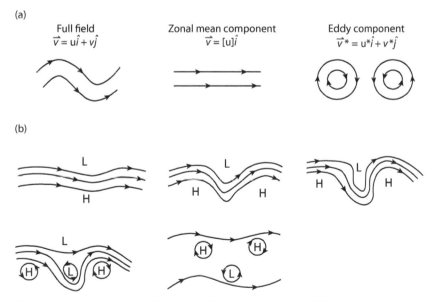

Figure 7.7. (a) Schematic illustrating the decomposition of the flow into zonal mean and eddy components. (b) The index cycle.

air to the north and warmer air to the south. Then, small eddy perturbations (waves) develop in the zonal flow initiated, for example, by uneven heating at the surface or within the atmosphere. In midlatitudes, these perturbations are unstable, meaning that they grow with time as the energy of the zonal mean flow is converted into eddy energy. The instability mechanism is known as *baroclinic instability*, and it occurs because of the strong meridional temperature gradients that are typical of middle latitudes (Figs. 2.6, 2.7, and 2.9). The waves continue to grow, and they "break" to form cutoff highs and lows. As depicted in Figure 7.7b, at the end of the index cycle warmer air has moved poleward and cooler air has moved equatorward, both contributing to the poleward transport of heat (and momentum) to balance the radiation deficit/surplus (Fig. 5.9). The index cycle is more active during the winter, when meridional temperature gradients are larger.

7.3 EXERCISES

7.1. Verify that Eqs. 7.5 and 7.6 satisfy Eq. 7.7.

7.2. State one way in which the Hadley and Walker circulations are similar, and one way in which they are different.

7.3. Consider a layer of the atmosphere extending from the surface to 800 hPa located on the equator. Within this layer, at what rate is the Hadley circulation transporting mass, and in what direction is this mass flux during boreal winter? During austral winter?

OCEAN CIRCULATION SYSTEMS

8

There are two fundamental drivers of ocean circulation systems: surface wind and density variations.

8.1 THE WIND-DRIVEN CIRCULATION: EKMAN DYNAMICS

Surface ocean currents are driven by surface winds, which transfer momentum into the ocean through frictional acceleration. The uppermost layers of the ocean then drag along the layers below. This frictional acceleration weakens with increasing depth, and the depth at which it is zero is the base of the ocean mixed layer. The essential physics of this process was developed by Swedish oceanographer V.W. Ekman, and the depth over which wind-driven ocean currents are generated is known as the *Ekman layer* in his honor.

Picture a quiescent initial state in the Northern Hemisphere, with no wind and an ocean surface at rest. Now imagine that a westerly wind begins to blow and it activates the surface waters. As soon as a parcel of water begins to move eastward ($u > 0$) under this frictional acceleration, it attains a southward (negative) meridional acceleration component due to the Coriolis acceleration, $dv/dt \sim - fu < 0$. As a result, the ocean surface current flows to the right of the surface wind's direction for this Northern Hemisphere example ($f > 0$), as illustrated in Figure 8.1. In the Southern Hemisphere ($f < 0$), the surface current is directed to the left of the surface wind.

Now, surface waters drag along the water immediately below, and the velocity of this lower layer is directed to the right of the surface water in the Northern Hemisphere, and to the left in the Southern Hemisphere, due to Coriolis acceleration. As deeper layers are accelerated by the layers above them,

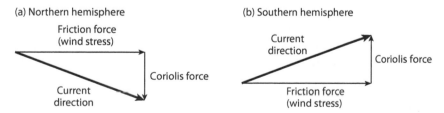

Figure 8.1. Directions of surface currents relative to surface winds in the (a) Northern and (b) Southern Hemispheres.

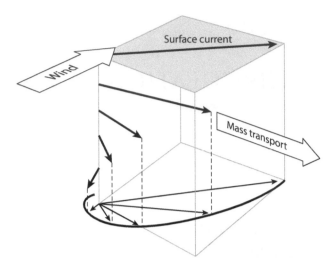

Figure 8.2 The Ekman spiral in the Northern Hemisphere.

the direction of the current spirals with increasing depth. This phenomenon is called the *Ekman spiral*. For a westerly surface wind, the Ekman spiral is clockwise in the Northern Hemisphere and counterclockwise in the Southern Hemisphere, with the current speed decreasing with depth as the frictional acceleration that originates at the surface weakens (Figure 8.2).

We can calculate current velocities using the horizontal momentum equations (Eqs. 6.56 and 6.57), simplified to capture the essential physics of the problem. Since mixed-layer currents are observed to occur under steady-state conditions and in the absence of advection, let

$$\frac{\partial u}{\partial t} = \frac{\partial v}{\partial t} = -\vec{v} \cdot \nabla u = -\vec{v} \cdot \nabla v = 0. \tag{8.1}$$

Wind-driven surface currents are also generated in the absence of pressure gradient forces within the ocean mixed layer, so the pressure gradient force can also be assumed to be zero, and the equations of motion reduce simply to

$$0 = f v_E + F_F^x \tag{8.2}$$

and

$$0 = -f u_E + F_F^y, \tag{8.3}$$

where the subscript E denotes the Ekman velocity components.

We now develop an expression for friction that represents the force of the wind on surface waters and the force of adjacent layers of water on each other. Define horizontal *stress*, $\vec{\tau} = \tau^x \hat{i} + \tau^y \hat{j}$, as the frictional force per unit area exerted on a parcel of water. A parcel at the surface experiences wind stress on its upper surface [$\tau^x(z_2) > 0$ when the surface wind is westerly] and frictional drag from the water below on its lower surface [$\tau^x(z_1) < 0$], as diagrammed in Figure 8.3. The geometry is the same for parcels below that are frictionally accelerated by the water layer above and decelerated by frictional drag from below. Let

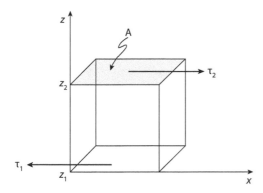

Figure 8.3. Frictional stresses on a parcel of air or water.

$\Delta \tau_x = \tau^x(z_2) - \tau^x(z_1)$ be the net stress exerted on a parcel of water with density ρ_W, thickness $\Delta Z = Z_2 - Z_1$, and top and bottom surface area A. In Eqs. 8.2 and 8.3, F_F^x and F_F^y are defined as forces per unit mass, so

$$F_F^x \sim \frac{\text{force}}{\text{mass}} \sim \frac{\text{force}}{\text{area}}\frac{\text{area}}{\text{mass}} \sim \Delta \tau^x \frac{A}{\rho_W \Delta z A} \rightarrow \frac{1}{\rho_W}\frac{\partial \tau^x}{\partial z}. \tag{8.4}$$

Similarly,

$$F_F^y = \frac{1}{\rho_W}\frac{\partial \tau^y}{\partial z}. \tag{8.5}$$

With these expressions for friction, the equations of motion (Eqs. 8.2 and 8.3) yield Ekman current velocities of

$$v_E = -\frac{1}{\rho_W f}\frac{\partial \tau^x}{\partial z} \tag{8.6}$$

and

$$u_E = \frac{1}{\rho_W f}\frac{\partial \tau^y}{\partial z}. \tag{8.7}$$

Equations 8.6 and 8.7 indicate that the horizontal components of the Ekman current velocity (u_E, v_E) depend on the vertical structure of the frictional stress. If we assume that density is constant through the mixed layer, these expressions for u_E and v_E can easily be integrated through the depth of the mixed layer to find the vertically integrated Ekman velocity, $\vec{V}$, defined by

$$\vec{V} \equiv \int_{-h}^{0} \vec{v}_E dz = \int_{-h}^{0} u_E dz \hat{i} + \int_{-h}^{0} v_E dz \hat{j} \equiv U\hat{i} + V\hat{j}, \tag{8.8}$$

where h is the mixed-layer depth. Performing the integration from the bottom of the Ekman layer ($z = -h$, $\tau^x = 0$, $\tau^y = 0$) to the ocean surface ($z = 0$, $\vec{\tau} = \vec{\tau}_S = \tau_S^x \hat{i} + \tau_S^y \hat{j}$), with the assumption that density is constant, we obtain

$$U = \int_{-h}^{0} \frac{1}{\rho_W f}\frac{\partial \tau^y}{\partial z} dz = \frac{\tau_S^y}{\rho_W f}, \tag{8.9}$$

$$V = -\int_{-b}^{0} \frac{1}{\rho_w f} \frac{\partial \tau^x}{\partial z} dz = -\frac{\tau_s^x}{\rho_w f}, \tag{8.10}$$

and

$$\vec{V} = \frac{1}{\rho_w f}\left(\tau_s^y \hat{i} - \tau_s^x \hat{j}\right) = -\frac{\hat{k} \times \vec{\tau}_s}{\rho_w f}. \tag{8.11}$$

The units of $\vec{V}$ are m²/s, so multiplying V times a horizontal length scale, L, yields a volume transport (m³/s) that measures the number of cubic meters of water transported each second by the *Ekman transport*. This flux is often expressed in sverdrup (Sv) units, where 1 Sv = 10^6 m³/s. Since the Ekman transport is proportional to the curl of the surface wind stress (Eq. 8.11), it is perpendicular the surface wind. In the Northern Hemisphere, the Ekman transport is to the right of the wind direction (Fig 8.2), and in the Southern Hemisphere it is to the left.

Recall from Figure 2.22 that ocean surface currents are organized into gyres, or anticyclonic circuits in each ocean basin, with westerly flow in middle latitudes, easterly flow in the tropics, cool (equatorward) eastern boundary currents, and warm (poleward) western boundary currents. Because the Ekman transport is perpendicular to the surface winds, water within the Ekman layer moves toward the center of the ocean basins, creating hills of water in the subtropics. For example, the Sargasso Sea in the tropical North Atlantic is some 1.5 m higher in the center than near the coasts. This redistribution of mass within the ocean basin sets up pressure gradients forces. A parcel of water traveling, for example, northward in the Gulf Stream experiences a westward pressure gradient force, away from the mass accumulation in the center of the North Atlantic basin, and an eastward Coriolis force.

The mounding of water in the ocean basins does not occur exactly in the center of the basins but, rather, in the western portion of the basin due to the eastward rotation of the earth. As a result, zonal pressure gradients in the western ocean basins are greater than those in the eastern basin. This *western boundary intensification* produces western boundary currents that are stronger, narrower, and deeper than eastern boundary currents. For example, in the Atlantic, peak Gulf Stream velocities exceed 2 m/s, while the Canary Current flow tops out at about 0.3 m/s. The Canary Current is characteristically 1000 km wide and 500 m deep, compared with the Gulf Stream, which is only 100–200 km wide but up to 2000 m deep.

This geostrophic argument provides a first-order understanding of western boundary intensification, but it does not completely explain observed western current velocities. The role of coastal friction in the balance of forces is also important.

Because Coriolis forces play a prominent role in the Ekman dynamics, the governing equations break down close to the equator. Note, for example, that Eqs. 8.6 and 8.7 yield infinite current velocities for $f = 0$. On the equator, currents are accelerated by pressure gradient forces that are directed eastward down the geopotential height gradients formed by the massing of water in the western parts of the ocean basin by the North and South Equatorial surface currents. The result is eastward-flowing *equatorial countercurrents*, which bring water mass back to the eastern ocean basins below the surface (~100 m) as *undercurrents*. These currents are sometimes described as jets because they tend to be narrow (~200 m) and fast (~1.5 m/s).

8.2   The Density-Driven Circulation: The Thermohaline Circulation

Away from the surface, the thermohaline circulation (Fig. 2.23) is driven by density differences related to temperature (Fig. 2.17) and salinity (Fig. 2.21) distributions. The circulation is slow and time scales are long—1000 years and longer for the global-scale ocean circulation.

The word's oceans are vast and mostly far from human habitation, making them difficult to observe. To supplement direct measurements, *transient tracers* are used to infer large-scale circulation in the ocean. These are substances that are introduced at the ocean surface, transferred downward, and circulated around the globe in a process known as *ocean ventilation*. Some of these tracers are of human origin and others are naturally occurring.

The testing of atomic bombs in the earth's upper atmosphere began in the 1950s and persisted until the Test Ban Treaty of 1963 prohibited the detonation of nuclear weapons in outer space, underwater, or in the atmosphere. These detonations produced an anthropogenic source of tritium (^{3}H), an isotope of hydrogen with a half-life of 12.45 years, that is much larger than the natural source due to the bombardment of molecular hydrogen by cosmic rays. Because tritium is chemically and biologically nonreactive in the atmosphere and oceans, and does not modify the circulation, it serves as a passive tracer of the circulation.

In the early 1970s, oceanographic cruises organized to measure tritium distributions found that ^{3}H had penetrated to greater depths at higher latitudes. Whereas the tritium was confined above about 500 m in the tropics and subtropics, it had mixed down to about 2 km in middle latitudes of the Northern Hemisphere. In the North Atlantic Ocean, high tritium levels were found down to the seafloor at about 4 km. Note that this tritium distribution is similar to the thermal structure, since North Atlantic surface waters are at about the same temperature as Atlantic deep waters (Fig. 2.17). Both distributions are evidence of the North Atlantic deep water formation discussed in section 2.2.

The Gulf Stream carries the warm, high-salinity water of the subtropical Atlantic (Fig. 2.19) to higher latitudes. The North Atlantic Ocean basin opens into the Labrador Sea and the Arctic Ocean, so the high-salinity flow continues north as the North Atlantic Drift (Fig. 2.22). Exposure to the cold, dry atmosphere at high latitudes drives strong longwave, sensible, and latent heat fluxes from the surface water (Figs. 5.11a and 5.12). The longwave back radiation is small due to low levels of moisture in the Arctic atmosphere (Fig. 5.11b), so the waters cool rapidly to 1.0°C–2.5°C. This strong cooling, combined with high salinity, produces high density and sinking.

Changes in the North Atlantic deep water formation influence the large-scale thermohaline circulation system. Decadal-scale variations in sinking rates, temperature, and salinity are documented, so it is clear that the system changes on time scales comparable to those of anthropogenic climate change. On longer time scales, stabilization (weakening) of the thermohaline circulation occurred at the end of the last glacial period when the melting of the great North American glaciers injected massive amounts of freshwater into the North Atlantic.

A naturally occurring transient tracer in the ocean is dissolved molecular oxygen. Surface waters become saturated with molecular oxygen due to absorption from the atmosphere and photosynthetic activity in the *euphotic zone*, or even supersaturated when ocean surface waves break. As these *young surface waters* move away from the atmosphere/ocean interface, that is, as the ocean is ventilated, oxygen is consumed in the oxidation of detritus and dissolved organic material. As a result, the level of oxygen saturation of a sample of ocean water is a measure of the *age* of the water, or how recently that water was at the surface.

Figure 8.4a shows a north/south cross section of oxygen saturation values in the Atlantic Ocean at 34.5°W longitude (to avoid land). Consistent with the inferences from tritium and temperature distributions, young water with saturation levels greater than 80% penetrates to the ocean bottom in the North Atlantic. The water with lowest oxygen saturation levels in the Atlantic, known as *old water*, is not located in the deep ocean but, rather, at a depth of about 500 m in the tropics.

The distribution of dissolved oxygen and, therefore, the large-scale circulation is somewhat different in the Pacific. As seen in Figure 8.4b, the oldest Pacific basin waters are found between 500 m and 2500 m depth in middle northern latitudes. These are the oldest waters in the ocean, with oxygen saturation levels below 20% and an estimated elapsed time since contact with the surface of 1000 years or more. There is no deep water formation at high northern latitudes in the Pacific due to geography, since the basin extends only to about 60°N.

The formation of *bottom water* along the southern shores of Antarctica is indicated in Figure 8.4b. The Antarctic bottom water flows north (see Fig. 2.23), beneath the deep water formed by sinking in the North Atlantic and guided by topography, with temperatures from 0°C to −1°C and salinity of about 34.7 psu. When sea ice forms and expands the ice shelves of the Weddell and Ross Seas in winter, the salinity of the underlying water is increased because most of the salt is not incorporated into the ice. This process is called *brine exclusion* or *brine rejection*. The resulting increase in sea water density drives the Antarctic bottom water formation. Another mechanism that supports the bottom water formation is cooling of surface waters when areas of open water, known as *polynyas,* form in the sea ice along the Antarctic coast. *Katabatic winds* blowing from the continent keep polynyas open by pushing the constantly forming sea ice to the south. Because the open surface waters have high salinity (due to brine exclusion) and low temperatures, they sink to the ocean bottom.

8.3 VERTICAL MIXING PROCESSES

So far we have discussed two types of large-scale ocean circulation systems, namely, the wind-driven currents of the mixed layer and the density-driven circulation of the global ocean. The thermohaline circulation intersects the

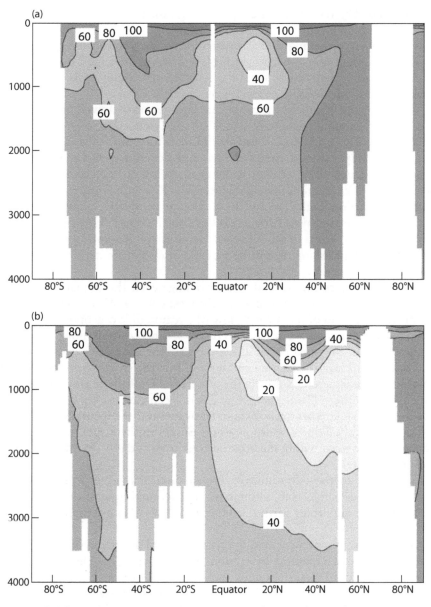

Figure 8.4 Annual percent oxygen saturation in the (a) Atlantic along 34.5°N, and (b) Pacific along 180°W. Contour intervals are 20% and the vertical axis is in m. Data from the National Oceanographic Data Center World Atlas (2005).

surface at very high latitudes, but how do the upper and lower layers of the ocean interact in the rest of the ocean? What processes lead to vertical motion and mixing across the ocean's layers?

One large-scale vertical mixing process is known as *Ekman pumping*. As discussed in section 8.1, surface currents within each ocean basin form anticyclonic gyres and the associated Ekman transport mounds water into the central ocean basins. The resulting vertical pressure gradients drive water out the bottom of the mixed layer.

Wind-driven upwelling and downwelling occur on smaller space scales. Coastal upwelling occurs when winds blow directly from land to ocean, driving warm surface water away from the coast. Cooler water wells up to replace it, obeying the constraint of conservation of mass.

Wind blowing parallel to a coast can also generate upwelling and downwelling. For example, southerly winds off the west coast of tropical South America (see Fig. 2.13) drive westward Ekman transport of water within the ocean mixed layer off the coasts of Peru and Ecuador. The water mass is replaced by the upwelling of cool water from below as well as the cool Peru Current (Fig. 2.22), and both help maintain the low temperatures in the Pacific cold-tongue region (Figs. 2.15 and 2.16). Similar wind-driven upwelling occurs off the California and West African coasts in the Northern Hemisphere and off the west coast of Africa in the Southern Hemisphere. Regions of upwelling are typically excellent fishing regions because the upwelling waters carry nutrients from the deeper ocean.

Ekman dynamics can generate downwelling when the Ekman transport is directed toward the coast, as occurs off the west coast of Alaska where the wintertime development of the Aleutian low (the region of low geopotential heights in the North Pacific shown in Fig. 2.4) places southerly winds along the coast.

Upwelling in the open ocean along the equator is another consequence of the Ekman dynamics. Recall that surface winds in the tropics converge near the equator, with northeasterly flow in the Northern Hemisphere and southeasterly flow in the Southern Hemisphere (Fig. 2.13). According to Eq. 8.11, these surface winds drive southeasterly Ekman transport in the Northern Hemisphere and northeasterly Ekman transport in the Southern Hemisphere. The result is a divergence of mass within the ocean mixed layer and *equatorial upwelling* of cool ocean water.

In addition to these dynamically driven vertical mixing processes related to the Ekman dynamics and, ultimately, the frictional drag of the wind on the ocean surface, important vertical mixing processes also are driven by density differences.

Salt fingering is a vertical mixing process related to the physics of *diffusion*, or the transport of a quantity by random molecular motion. Heat and momentum can be diffused, as well as mass such as salt. Different quantities diffuse at different rates. When a fluid's density is determined by two different components or properties with different rates of diffusion, *double-diffusive processes* can occur and contribute to vertical mixing. Salt fingering is one example of double-diffusive convection.

To understand the salt-fingering process, consider a tank of cold freshwater with an embedded region of warm salty water (Fig. 8.5). The warm salty region is not physically mixed or stirred into the freshwater, but both heat and salt will diffuse into the cool freshwater. Heat diffusion takes place more quickly than salt diffusion, so the tank will become isothermal before it attains uniform salinity.

Now, consider the two-layer fluid system drawn in Figure 8.6a. The top fluid has a lower density than the bottom fluid, so the system is hydrodynamically

Figure 8.5 A region of warm salty water in a tank of cooler, fresh water.

(a)   (b)

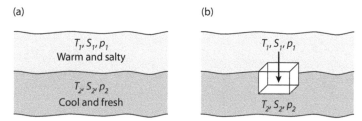

Figure 8.6 (a) A two-layer ocean system with warm salty water overlying cold freshwater. (b) A perturbation of the two-layer ocean system.

stable. Note that temperature dominates in the determination of density in this idealized problem, but this is the case in most of the ocean (high latitudes excepted).

Next, test the stability of the two-layer system by moving a parcel of water from the upper layer to the lower layer (Fig. 8.6b). The system is stable, and if we make the adiabatic assumption, the less dense, displaced parcel will float back to the surface. More realistically, however, the parcel will tend to lose heat quickly through diffusion, especially if the displacement is slow or the rebound is delayed. Salt will also diffuse out of the displaced parcel, but at a much slower rate. Therefore, the density of the displaced parcel will increase as it cools (due to heat diffusion). If the temperature of the parcel equilibrates with that of the bottom layer but the parcel retains its higher salinity, it will be denser than the fluid in the bottom layer and continue to sink. Thus, double-diffusive convection makes this apparently stable two-layer system unstable.

Salt fingering occurs in the ocean where salty, warm waters overlie fresher, cool waters. Subtropical regions, such as the Mediterranean Sea, are favorable for this process since the excess of evaporation over precipitation increases the salinity of surface waters. Parts of the Atlantic basin have warm, salty surface waters over cooler freshwater due to the outflow of warm, salty Mediterranean water.

Caballing is a vertical mixing process that occurs because the equation of state for ocean water is nonlinear. Counterintuitively, mixing together two parcels of saltwater with the same density can result in a parcel with a density greater than that of the original parcels.

Consider the example of a layer of water beneath an ice shelf. The water is at 0°C and has a salinity of 34 psu. It overlies warmer but saltier water at 10°C

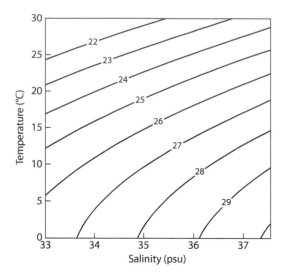

Figure 8.7 A *T–S* plot for ranges in temperature and salinity (psu) relevant to the oceans, at surface pressure.

and 36.5 psu. According to the *T–S* plot shown in Figure 8.7, which is similar to Figure 5.1 but restricted to salinities typical of the ocean, the top layer of water is less dense ($\sigma_t = 27.3$) than the lower layer ($\sigma_t = 28$). This is a stable configuration. If some perturbation brings a parcel of water from the upper layer into the lower layer, the parcel will float back to the upper layer, and the initial state will be reestablished.

Now, imagine that the ice thickens, and the water in the layer beneath the ice becomes more saline due to brine exclusion. Suppose that the salinity of the top water layer increases to 34.8 psu. This water is still fresher than that in the lower layer but its density has increased to $\sigma_t = 28$, the same as that of the lower layer. Having the same density, the two layers are now free to mix. Assuming that the masses of the two layers are equal, the temperature and salinity of the mixed water will be the average of the two layers, 5°C and 35.25 psu, respectively. According to the *T–S* plot in Figure 8.7, the density of the mixture is $\sigma_t = 28.2$, which is greater than the density of the unmixed layers. The newly mixed water will sink.

8.4 REFERENCE

Garcia, H. E., R. A. Locarnini, T. P. Boyer, and J. J. Antonov, 2006. World Ocean Atlas 2005. S. Levitus, ed. NOAA Atlas NEDIS 63, U. S. Government Printing Office, Washington, DC.

8.5 EXERCISES

8.1. Suppose that a wind from the west produces a stress on the ocean at 45°N latitude. In what direction is the Ekman transport? If the stress is 0.1 Pa, what is the magnitude of the Ekman volume transport across a 1000 km line? Express your answer in cubic meters per second (m³/s) and in sverdrups (Sv).

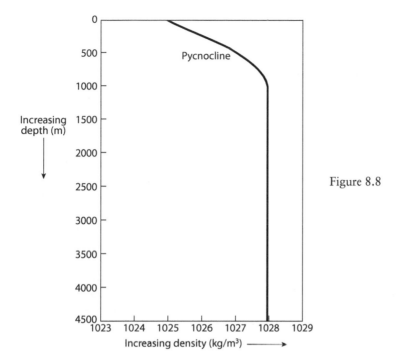

Figure 8.8

8.2. As it is for the atmosphere, hydrostatic balance is an excellent
approximation for the ocean. Integrate the hydrostatic equation to find the
pressure at the base of the pycnocline at about 1 km depth in the ocean.
This integration is different from the atmospheric case because you cannot
use the ideal gas law to eliminate density. Use Figure 8.8 to find a way to
treat density in this problem, and consult figures in previous chapters for
any values you need to complete the calculation. Make sure your results are
reasonable, and justify all assumptions.

9
THE HYDROLOGIC CYCLE

The global hydrologic cycle links the climate system components (Figs. 1.1 and 1.2), with water constantly moving through the world's oceans, the cryosphere, the lithosphere (in lakes, rivers, and soils), the atmosphere, and the biosphere. Figure 9.1 provides a schematic overview of these interactions and some of the processes that must be considered to understand the hydrologic cycle.

The hydrologic cycle can be evaluated by constructing budgets based on the concept of conservation of (water) mass, similar to the heat balance equations written in chapter 5. Budgets for an atmospheric column and a land surface volume are derived in the following sections. In general, observations of global and regional hydrology are not sufficiently accurate to allow us to quantify individual terms in the water budgets beyond fairly rough estimates, such as those shown in Figure 2.24.

9.1 ATMOSPHERIC WATER BALANCE

Although the mass of water in the atmosphere is tiny compared with the other reservoirs of water in the climate system (see Table 2.1), its radiative properties (chapter 4) make atmospheric water vapor a primary determining factor of climate on all space and time scales.

Consider a rectangular column of air with unit cross-sectional area extending from the surface to the top of the atmosphere (Fig. 9.2). Water vapor increases in the volume when water evaporates (E), and decreases when water vapor condenses and precipitation (P) removes water from the atmospheric column. The water vapor content can also change when the atmospheric circulation converges water vapor into the column (Fig. 9.2a) or diverges water vapor out of the column (Fig. 9.2b). This simple water vapor budget neglects ice formation and sublimation, and assumes that all liquid water falls to the surface. Note that because the physical surface is excluded from the atmospheric column, the redistribution of water on the surface by runoff, subsurface flow, or ocean currents is not included in the atmospheric budget.

To develop an equation that governs conservation of water vapor in an atmospheric column, we write a mathematical expression for the time rate of change of the amount of water vapor in the volume and set it equal to the sum of the sources and sinks of water for the volume. The total amount of water vapor in the column per unit cross-sectional area, W (kg-water/m^2), is calculated by integrating the elemental water vapor mass, dW, over the volume. If ρ_V is the density of water vapor in the volume, then

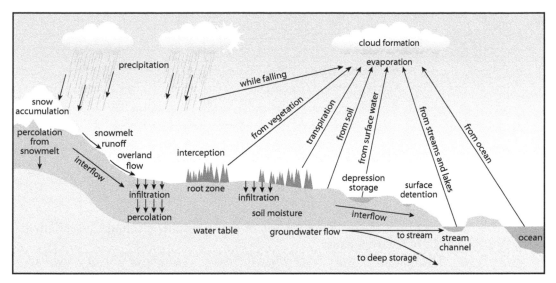

Figure 9.1 Schematic representation of exchanges and processes important for the hydrological cycle. From http://rst.gsfc.nasa.gov/Sect16/Sect16_4.html/.

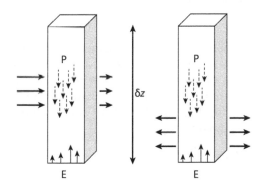

Figure 9.2 The atmospheric column moisture balance. Arrows indicate moisture transport.

$$dW = \rho_V \, dz. \tag{9.1}$$

Using the definition of specific humidity (chapter 2),

$$q = \frac{\rho_V}{\rho}, \tag{9.2}$$

where ρ is the density of the volume (including air and water vapor), we have

$$dW = \rho q \, dz. \tag{9.3}$$

Then,

$$W = \int_{0}^{z_{TOP}} dW = \int_{0}^{z_{TOP}} \rho q \, dz, \tag{9.4}$$

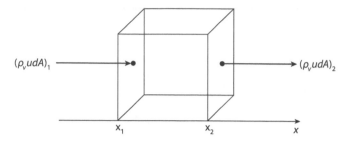

Figure 9.3 Zonal moisture transport through an elemental volume.

where z_{TOP} is the height of the top of the atmospheric column, essentially the height of the tropopause.

The value of W will change when there is a net source or sink of water vapor:

$$\frac{dW}{dt} = \Sigma(\text{sources} - \text{sinks}). \qquad (9.5)$$

Sources include evaporation from the surface and from liquid water suspended in or falling through the column, E, and moisture convergence into the column by the atmospheric circulation. Sinks of water vapor are precipitation, P, and moisture divergence by the circulation.

Consider moisture convergence in the x-direction for the elemental volume drawn in Figure 9.3. The rate at which water mass is carried into the volume by the zonal flow *at* x_1 through an area dA perpendicular to the wind vector is $(\rho_v u dA)_1$, which has units of kg_{H_2O}/s. Similarly, the rate at which water is transported out of the volume is $(\rho_v u dA)_2$. Using Eq. 9.2, the rate of water vapor mass accumulation in the volume due to zonal transport is

$$\left[(\rho q u)_1 - (\rho q u)_2\right]dA, \qquad (9.6)$$

and the rate of water vapor mass accumulation per unit volume associated with the zonal flow is

$$\frac{\left[(\rho q u)_1 - (\rho q u)_2\right]dA}{dV} = -\frac{\left[(\rho q u)_2 - (\rho q u)_1\right]dA}{dA(x_2 - x_1)} \quad \rightarrow \quad -\frac{\partial(\rho q u)}{\partial x}. \qquad (9.7)$$

Note that if $(\rho q u)_1 > (\rho q u)_2$, then $\partial(\rho q u)/\partial x < 0$, and water vapor is converging in the volume. Generalizing to three dimensions, the rate of water vapor mass accumulation in the volume is

$$-\frac{\partial(\rho q u)}{\partial x} - \frac{\partial(\rho q v)}{\partial y} - \frac{\partial(\rho q w)}{\partial z} \equiv -\nabla \cdot (\rho q \vec{v}). \qquad (9.8)$$

The term on the right-hand side of (Eq. 9.8) is called the *water vapor moisture flux convergence*.

Combining Eqs. (9.4), (9.5), and (9.8), we obtain the equation for the atmospheric column water vapor balance:

$$\frac{dW}{dt} = \frac{d}{dt}\left(\int_0^{z_{TOP}} \rho q \, dz\right) = -\int_0^{z_{TOP}} \nabla \cdot (\rho q \vec{v}) \, dz + E - P. \tag{9.9}$$

For the climatology, it is reasonable to assume that water vapor does not accumulate in the atmospheric column $(dW/dt) = 0$ and Eq. 9.9 becomes

$$P - E = -\int_0^{z_{TOP}} \nabla_h \cdot (\rho q \vec{v}) \, dz. \tag{9.10}$$

Review the precipitation and surface evaporation climatologies presented in chapter 2 (Figs. 2.25–2.29), and note again that their geographic distributions are quite different so that, locally, $P \neq E$. This dissimilarity means that the redistribution of moisture in the atmosphere by the circulation plays an important role in determining the precipitation distribution.

9.2 LAND SURFACE WATER BALANCE

An understanding of the processes that control the availability of water on the land surface is important for agricultural applications, assessments of water resources, and prediction of floods and droughts. The rate at which water accumulates in a volume of the land surface must account for input from precipitation, and loss from evapotranspiration and runoff. Mathematically,

$$S = P - E - R_s - R_u, \tag{9.11}$$

where

S = rate of storage of water (change in water content with time)
P = precipitation rate
E = evapotranspiration rate
R_s = surface runoff
R_u = underground runoff.

In Eq. 9.11, P includes water delivered to the surface in the form of rain, snow, and ice. Evapotranspiration includes evaporation from a water surface or bare land, *transpiration*, defined as the release of water vapor from plants, and *sublimation*, defined as the release of water vapor from ice and snow.

Flooding may occur when S is positive, that is, when the amount of water on the surface increases with time, and droughts are related to persistent negative values of S. Intense rainfall events that deliver high precipitation rates that cannot be balanced by evaporation and runoff instigate most flooding events.

For the climatology (denoted by overbars), and averaging over relatively large regions (denoted by square brackets), the rates of water storage and underground runoff are small and the dominant terms in the surface water balance are

$$[\overline{E}] = [\overline{P}] - [\overline{R}_s]. \tag{9.12}$$

9.3 EXERCISES

9.1. What are the units of water vapor moisture flux convergence?

9.2. Use Figures 2.25 and 2.28 to quantify the role of the Hadley circulation in the water balance of an atmospheric column in (a) the tropics; (b) the subtropics.

9.3. Show that in midlatitudes, with geostrophic flow, Eq. 9.10 can be approximated as

$$P - E = -\int_0^{z_{TOP}} \rho \vec{v} \cdot \nabla_h q \, dz.$$

Note all assumptions you make in the approximation.

10

RADIATIVE FORCING OF CLIMATE CHANGE

The following three chapters are concerned with climate change. They build on the background presented in the previous chapters and discuss how and why climate changes, with a focus on contemporary climate change issues. We begin with the pure radiation side of the issue, examining current changes in the atmosphere's chemical composition and the effects of those changes on the flow of shortwave and longwave radiation through the atmosphere in the absence of any response of the climate system to the changes in radiation. In chapter 11, we focus on the climate system's response to these changes in radiative fluxes. Finally, we discuss climate change prediction using climate models in chapter 12.

Earth's climate has changed throughout its 4.5 billion year history in association with *radiative forcing factors*, including changes in insolation and in the chemical composition of the atmosphere. On the longest time scales, up to billions of years, changes in solar luminosity force climate change. The sun is evolving and is about halfway through its 10-billion-year life as a *main sequence* star, that is, a star fueled by the conversion of hydrogen into helium. Over the first 4.6 billion years of the sun's life its luminosity has increased by about 25% and, according to models of stellar evolution, it will continue to increase for another 5 billion years until all its hydrogen has been converted to helium.

The development of life on the planet, along with early periods of intense volcanic activity and outgassing, altered the chemical composition of the atmosphere and also contributed to climate change on time scales of millions and billions of years. Plate tectonic processes modify the land/sea distribution and build mountains and, as a result, climate adjusts on time scales of millions of years. On time scales of tens of thousands of years, changes in the earth's orbital parameters (see chapter 3) force the glacial–interglacial oscillation that has characterized climate during the Holocene (the most recent 1.2 million years).

And now, climate is changing due to human activity. Since the Industrial Revolution in about 1750, burning of fossil fuels (coal, oil, and natural gas) has released CO_2 into the atmosphere. A host of other atmospheric constituents are increasing as well, including powerful greenhouse gases such as methane and nitrous oxide. The atmosphere's particulate (aerosol) concentration is also changing, along with land surface features such as albedo and surface roughness due to altered plant distributions, paving, and buildings.

10.1 THE ATMOSPHERE'S CHANGING CHEMICAL COMPOSITION

Our concern about anthropogenic climate change stems from the observed changes in the concentrations of greenhouse gases in the atmosphere. Of particular importance are increasing levels of carbon dioxide (CO_2), methane (CH_4), nitrous oxide (N_2O), chlorofluorocarbons (CFCs), and tropospheric ozone (O_3).

CARBON DIOXIDE (CO_2)

Regular measurements of the atmospheric CO_2 concentration were established at the Mauna Loa Observatory (elev. 3400 m) in Hawaii in 1958. Measurements are obtained by determining the degree of absorption of an infrared beam as it passes through an air sample. This record, which is known as the *Keeling*[1] *curve*, is shown in Figure 10.1a. The black line represents annual mean CO_2 levels, while the gray curve indicates monthly mean values. The unit is parts per million (*ppm*) of CO_2 in the air by volume. A CO_2 concentration of 1 ppm means that in 10^6 (1 million) liters of air the CO_2 occupies a volume of 1 liter. Between 1958 and 2011, annual mean CO_2 levels increased about 25%, from 315 ppm to 392 ppm.

Seasonal oscillations of about 5 ppm are apparent in the monthly mean data. Because there is much more land and, therefore, vegetation in the Northern Hemisphere than in the Southern Hemisphere, CO_2 levels drop during Northern Hemisphere summer due to photosynthetic activity.

The Mauna Loa CO_2 measurements have proved to be invaluable for establishing incontrovertibly that atmospheric CO_2 levels are rising because the record is consistent and continuous. The Mauna Loa data reflect global CO_2 levels because CO_2 is well mixed in the atmosphere. Molecules that do not react chemically in the atmosphere, or do not rain out, remain in the atmosphere for long periods of time and become uniformly distributed across the globe. Characteristic mixing times for the atmosphere range from a few days (e.g., for gases mixing zonally in middle latitudes), to a few months (mixing between hemispheres), to a few years (mixing between the troposphere and stratosphere). *Residence time* measures the average length of time a given molecule released into the atmosphere will remain there. For CO_2, residence times range from 5 to 300 years. The range is wide because the removal processes, which include uptake by the oceans and photosynthesis, operate on disparate time scales, but even the shortest residence time is longer than the characteristic global mixing time. Thus, the CO_2 record from the South Pole, shown in Figure 10.1b for 1975–2011, is similar to the Mauna Loa record except for the greatly reduced amplitude of the seasonal signal.

---

[1] Charles D. Keeling (1928–2005), with support from Roger Revelle and Harry Wexler, originated and sustained precise measurements of atmospheric CO_2 concentrations at the Mauna Loa Observatory, Hawaii, beginning in 1958, to address the hypothesis that human activity is changing the chemical composition of the atmosphere.

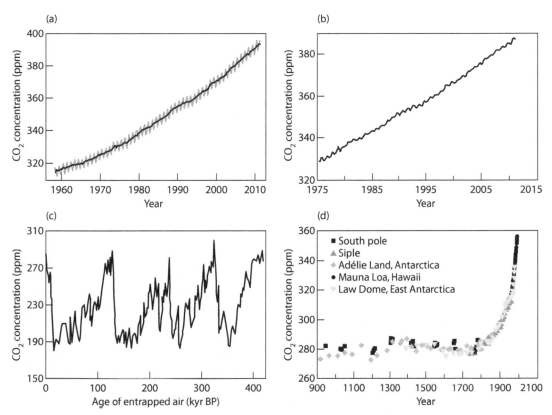

Figure 10.1 (a) Direct atmospheric CO₂ measurements from the (a) Mauna Loa Observatory and (b) South Pole. (c) Values for 400,000 years of atmospheric CO₂ from the Vostok, Antarctica, ice core. (d) Values of atmospheric CO₂ for the last 1100 years from various ice cores and Mauna Loa.

These observed changes in the atmospheric CO_2 concentration are known to be the result of human activity. Studies of isotopic ratios of atmospheric CO_2, as well as analyses of records of fossil fuel use and land clearing, prove that the added CO_2 in the atmospheric is of anthropogenic origin.[2]

The largest source of anthropogenic atmospheric CO_2 is the burning of fossil fuels, which converts organic carbon into CO_2 gas. This source is currently estimated at 6 Pg/yr.[3] The trend in rates of fossil fuel burning is similar to that of the atmospheric CO_2 concentration, that is, exponentially increasing, and the excess carbon in the atmospheric has an isotopic signature that confirms its origin. Additional sources of atmospheric CO_2 are forest clearing and other anthropogenic changes in the land surface (~1.0 Pg/yr) and cement manufacturing (~0.1 Pg/yr). Only about half the CO_2 released by human activity remains in the atmosphere; the rest is absorbed by the ocean (~1.9 Pg/yr) and

---

[2] The $^{13}C/^{12}C$ ratio of fossil-fuel CO_2 is about 2% lower than that of the pre-industrial atmosphere.

[3] The unit Pg is petagrams of carbon, or 10^{15} g of carbon; 1 Pg is equivalent to 1 Gt (gigaton), or 10^9 or 1 billion metric tons of carbon.

the terrestrial biosphere (~1.0 Pg/yr). The global carbon budget is discussed in more detail in chapter 12.

Ice cores from Vostok, Antarctica, provide a record of atmospheric CO_2 levels back more than 400,000 years. These values are based on the analysis of air bubbles trapped when ice forms. The data are highly reliable since the ice formation process essentially samples the air at the time of its formation and protects the sample. Similar records are available from the Arctic, but the Vostok ice cores are the longest (reaching a depth of nearly 4 km), having been taken from the very stable and deep East Antarctic ice. Figure 10.1c shows a record of atmospheric CO_2 concentration from Vostok. The most recent value recorded in the ice core is about 280 ppm, which is the pre-industrial CO_2 concentration. Recent increases (Figs. 10.1a and b) are so rapid they are not resolved in the ice cores. Proceeding backward in time, from left to right in Figure 10.1c, CO_2 levels fell below 200 ppm at the height of the last glacial period, about 20,000 years ago. CO_2 values remained below 250 ppm during the entire glacial period and rose to about 280 ppm during the last interglacial period, about 130,000 years ago. CO_2 levels have oscillated with the coming and going of glacial periods, increasing to about 280 ppm during interglacials and falling to about 200 ppm during glacial periods. This glacial–interglacial cycling is ultimately forced by well-known changes in the earth's orbital parameters (see section 3.5). The CO_2 oscillation amplifies the externally forced temperature changes and occurs primarily because a warmer ocean releases dissolved CO_2 in the same way that soda in an open glass goes flat when it is allowed to warm to room temperature.

Following the most recent glacial period, which ended about 20,000 years ago, atmospheric CO_2 concentrations remained at the background interglacial value of about 280 ppm until the late 1700s. Figure 10.1d combines measurements from Mauna Loa (after 1958) with information gathered from ice cores at various Antarctic stations. These data show dramatically that the current increases in atmospheric CO_2 are unprecedented and exponential.

METHANE (CH_4)

Methane is a powerful greenhouse gas with a more complicated story than CO_2. It has numerous sources and sinks, is not well mixed globally, and undergoes chemical reactions in the atmosphere. The largest sources of atmospheric CH_4 are emissions from wetlands and rice paddies. Other important sources are animals, termites, biomass burning, landfills, and fossil fuel (coal and gas) production. The primary sink for atmospheric CH_4 is its photochemical oxidation by the hydroxyl radical (OH) in the troposphere (see chapter 12). As a result, the atmospheric residence time for CH_4 is about 12 years, much shorter than for CO_2.

Direct measurements of atmospheric CH_4 since 1987 are shown in Figure 10.2a for stations at Mauna Loa and Barrow, Alaska. The fact that these values are different indicates that CH_4 is not well mixed in the atmosphere, but at both locations CH_4 increased steadily from 1987 until about 2000, stabilized from about 2000 to 2007, and then resumed rising.

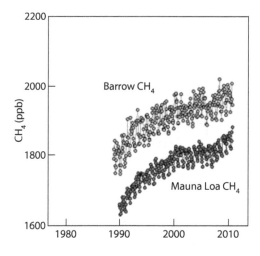

Figure 10.2 Methane concentrations measured at Barrow, Alaska (gray), and Mauna Loa, Hawaii (black) in ppb (parts per billion).

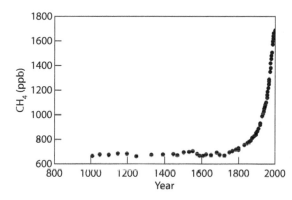

Figure 10.3 Methane concentrations from the Antarctic and Greenland ice cores.

Methane measurements are extracted from ice cores along with those of CO_2, and the Vostok core indicates that CH_4 varies with the glacial–interglacial oscillation in the same way as CO_2. Figure 10.3 shows CH_4 concentrations on the millennial time scale from Greenland and Antarctic ice core measurements. The pre-industrial CH_4 concentration was about 700 ppb (parts per billion) and it is currently increasing exponentially at an average rate of about 1% per year. This rise began in the 1700s as agricultural practices became more intense and widespread and, later, as industrial activity developed.

NITROUS OXIDE (N_2O)

Nitrous oxide, also known as "laughing gas" when used in high concentrations as a sedative in the dentist's office, is a powerful greenhouse gas with a long residence time in the troposphere of more than 100 years. Its concentrations are estimated to be increasing at a rate of 0.26% per year (see Fig. 10.4). Pre-industrial values were about 260 ppb and they had risen to about 324 ppb in 2010.

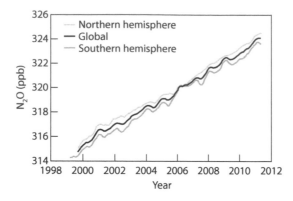

Figure 10.4 Global and hemispheric trends in concentration of atmospheric N_2O estimated from direct measurements.

The global nitrogen budget is studied to understand atmospheric N_2O values. Nitrous oxide is released from many different sources due to biological activity in soils and water—for example, microbial activity in tropical forests. Important anthropogenic sources are related to soil and manure management in agricultural practice, sewage treatment, and fossil fuel combustion. See chapter 12 for more details.

TROPOSPHERIC OZONE (O_3)

Ozone in the stratosphere absorbs ultraviolet solar radiation (see chapter 4), protecting life on the planet from these destructive wavelengths. But in the troposphere, O_3 is a pollutant that can damage living tissue, and it is a greenhouse gas.

The largest natural source of tropospheric ozone is downward migration from the stratosphere; lightning also generates O_3 by breaking down molecular oxygen. Ozone is produced by a host of human activities, including biomass burning, fossil fuel combustion, and the use of chemical solvents. Even photocopying and laser printing produce O_3. These activities release nitrogen oxides (NO_x, or NO and NO_2, not to be confused with N_2O), carbon monoxide (CO), and *volatile organic compounds* (VOCs) into the atmosphere, where they support the photochemical production of O_3. Ozone breaks down in sunlight, and one product of this photodissociation is the highly reactive OH radical, which helps remove CH_4 from the atmosphere. Uptake by plants is also an important sink.

Tropospheric O_3 varies significantly on all space and time scales, so producing an accurate measurement of its average levels would require an extensive observing network. Because such a network does not exist, global O_3 levels are estimated using computer models, and the validity of these models is tested against the relatively few measurements available. The most recent estimates take into account increases in tropospheric O_3 due to its anthropogenic emission as well as from compounds that react in the atmosphere to produce O_3. Negative tendencies in tropospheric O_3 due to the depleted ozone source in the stratosphere are also included in the estimates. Locally, in situ ozone measurements can be used to evaluate trends. A consistency among direct measurements of tropospheric ozone levels has not been established; some local measurements show increases at rates as high as 1.4% per year, while measurements in other areas find steady values.

CHLOROFLUOROCARBONS AND HALONS

Chlorofluorocarbons (CFCs) contain atoms of carbon, fluorine, and chlorine. They are manufactured for use in spray cans, blowing foams and packing material, and as solvents and refrigerants. In their use as refrigerants they are designed to absorb longwave radiation, so it is no surprise that they are extremely strong greenhouse gases. CFCs are nonreactive in the troposphere, which contributes to their long residence times, and when they migrate into the stratosphere they are decomposed by ultraviolet light and contribute to the destruction of stratospheric ozone.

The Montreal Protocol of 1987, strengthened by the London Agreement in 1990, established regulations to reduce CFC emissions to protect stratospheric ozone. Substitutes for these ozone-destroying compounds have been developed, for example, HFCs (hydrofluorocarbons) and HCFCs (hydrochlorofluorocarbons). These compounds react with OH and, therefore, have shorter atmospheric lifetimes and less migration into the stratosphere, but they are still strong greenhouse gases. Figure 10.5 shows the trends in the concentrations of selected CFCs and HCFCs in the troposphere.

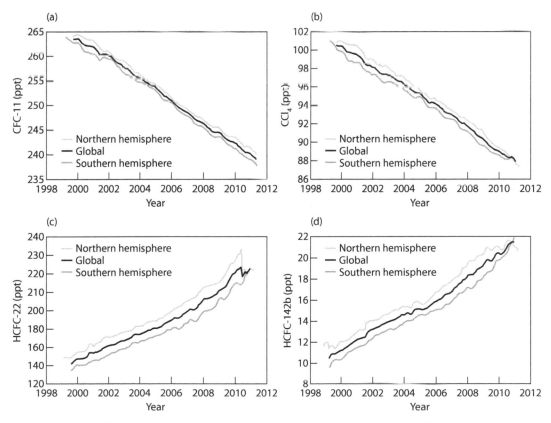

Figure 10.5 Global concentrations of (a) CFC-11, (b) CCl_4, (c) HCFC-22, and (d) HCFC-142b. Data from NOAA/ESRL halocarbons program. Units are parts per trillion (ppt).

Various other fluorine-containing gases are manufactured and used in industrial processes, sometimes as substitutes for ozone-depleting CFCs and HCFCs. Their emission rates are small, but they are extremely potent greenhouse gases. An example is sulfur hexafluoride (SF_6), one of the most powerful greenhouse gases (by mass). It is very long-lived in the atmosphere and is increasing at a rate of about 0.2 ppt (parts per trillion) per year, or about 7% per year. Sulfur hexafluoride is released into the atmosphere primarily by electric power industries.

Halons, or bromofluorocarbons, are fluorocarbons that contain bromine or iodine. A common use of halons is as fire extinguishers, and they are fairly long-lived, powerful greenhouse gases.

AEROSOLS

Atmospheric aerosols are suspended solid particles and liquid droplets, ranging in diameter from a few nanometers (1 nm = 10^{-9} m) to tens of micrometers. Atmospheric aerosols include smoke, water droplets, sulfate particles, organic carbon, black carbon, mineral dust, nitrates, smog, and sea salt; hybrid aerosols form when they interact and/or clump together.

Several natural sources of aerosols significant for climate have been identified. *Volcanic aerosols* are most commonly formed when ejected sulfur dioxide is converted into droplets of sulfuric acid. *Mineral dust aerosols* are lifted by surface winds over deserts and dry regions, for example, during periods of drought. The Sahara Desert is the largest global source. Forest fires emit smoke, black carbon, and organic carbon. Plants release VOCs, primarily isoprene, which is converted, in part, to OH in the atmosphere and also contributes to tropospheric ozone production. Ocean microalgae release *dimethylsulfide (DMS)*, a precursor compound for sulfate aerosols.

Anthropogenic aerosols originate in a wide variety of human activities. The greatest source is sulfate aerosols from fossil fuel (both coal and oil) combustion. This anthropogenic source is currently thought to be greater than the natural sources of sulfate aerosols. Biomass burning, for example, to clear land is another prominent source of anthropogenic aerosols in the form of organic carbon and black carbon. VOCs are emitted by thousands of manufactured products, including paints, furnishings, pesticides, glues, and office equipment. Some are toxic to humans, and they are important contributors to indoor pollution.

Most aerosols in the troposphere have relatively short residence times—under a few weeks. Removal processes include both wet and dry deposition. Aerosols in the stratosphere have longer residence times—often a year or more. The removal of stratospheric aerosols is related to mass exchange between the stratosphere and troposphere and is thought to occur in intense weather systems and upper-tropospheric jet streams.

10.2 RADIATIVE EFFECTS OF GREENHOUSE GAS INCREASES

What are the implications of the observed changes in greenhouse gases for climate? The first step in answering that question is to understand how the radiative fluxes in the atmosphere are affected.

Two methods of expressing the influence of changes in the atmosphere's composition are regularly used. One is the *direct radiative forcing*, defined as the change in the net downward radiative flux at the tropopause due to a given change in an atmospheric constituent. In other words, the direct radiative forcing is the amount of radiative energy (W/m^2) added to the troposphere due to a given change in a greenhouse gas. In calculating the direct radiative forcing, all radiative processes and adjustments are included, in both the troposphere and the stratosphere and for both longwave and shortwave fluxes, but all climate parameters (e.g., atmospheric and surface temperatures, specific humidity) are fixed—that is, the climate response is not included.

The direct radiative forcing due to the observed changes in greenhouse gases since pre-industrial times is listed in Table 10.1. Human activity had increased atmospheric CO_2 concentrations by about 40% as of 2011, resulting in a direct radiative forcing of 1.66 W/m^2. The next largest contribution is from CH_4, with tropospheric O_3 close behind. Among the manufactured compounds, CFC-12 has produced the greatest perturbation to the climate system's heat balance.

A number of complications and nuances arise in calculating the overall radiative effects of changes in greenhouse gases. For example, absorption bands of atmospheric constituent gases can overlap and modify one another's absorption properties. CO_2 and H_2O, as well as N_2O and CH_4, for example, have some overlapping absorption bands.

Another measure used to quantify and compare the radiative effects of greenhouse gas emissions is the *global warming potential* (GWP). GWP evaluates the implications over time of releasing a unit mass of a given greenhouse gas compared with the release of a unit mass of CO_2. The calculation takes into account the strength of the greenhouse gas as well as its residence time in the atmosphere. As listed in Table 10.1, a 100-year time horizon is often used.

The GWPs of all the greenhouse gases of concern are significantly greater than 1, meaning that their potential ability to modify climate is much greater than that of CO_2 *per unit mass*. Some of this potential derives from the constituent's molecular structure, which allows the component to absorb more longwave radiation. For example, atmospheric CH_4 has lower concentrations and shorter residence times than CO_2, but its GWP is 25 times greater than that of CO_2 (see Table 10.1). N_2O has both a long residence time and strong absorption, and its GWP is nearly 300. But the GWPs of the manufactured compounds dwarf those of the naturally occurring gases. Sulfur hexafluoride has the highest GWP because it is a powerful greenhouse gas and also has a long residence time.

Effects of aerosols on climate are complicated, not completely known, and highly regional. Aerosols produce direct radiative forcing on climate due to both absorption and scattering. Increased aerosol loading can either cool or warm climate, depending on the type of aerosol and its distribution as well as on the albedo of the underlying surface. Cooling (negative direct radiative forcing) is induced over dark surfaces such as oceans or forests, and warming (positive direct radiative forcing) is induced over bright surfaces such as snow and deserts. To complicate matters further, aerosols exert an indirect effect on the earth's radiation budget when they interact with and modify clouds.

Calculations of the direct radiative forcing and GWP essentially convert a given greenhouse gas emission into an energy perturbation (W/m^2), allowing us

Table 10.1. Greenhouse gas concentrations and their radiative effects

Gas	Preindustrial tropospheric concentration	Current concentration	Direct radiative forcing (W/m^2)	GWP (100 yr)	Atmospheric residence time
Carbon dioxide (CO_2)	280 ppm	390 ppm	1.66	1	~100 yr
Methane (CH_4)	700 ppb	1,820–1,890 ppb	0.48	25	12 yr
Nitrous oxide (N_2O)	270 ppb	324	0.16	298	114 yr
Tropospheric ozone (O_3)	25 ppb	34	0.35	N/A	Hours to days
CFC-11	0	239	0.063	4,750	45 yr
CFC-12	0	527	0.17	10,900	100 yr
CF-113	0	77	0.024	6,130	85 yr
CCl_4	0	87	0.012	1,400	26 yr
HCFC-142b	0	20 ppt	0.0031	2,310	17.9 yr
HCFC-22	0	220 ppt	0.033	1,810	12 yr
Halon 1211	0	4.1 ppt	0.001	1,890	16 yr
SF_6	0	7.3 ppt	0.0029	22,800	3200 yr

Source: Table adapted from Recent Greenhouse Gas Concentrations by T. J. Blasing, DOI: 10.3334/CDIAC/atg.032, http://cdiac.ornl.gov/pns/current_ghg.html.

Note: Values for SF_6 are approximate since concentrations are always changing and the constituent may not be well mixed. Concentrations listed here are an average of levels reported for Mauna Loa and the South Pole for May 2011 by NOAA at http://www.esrl.noaa.gov/gmd/dv/hats/cats/cats_conc.html.

to compare one greenhouse gas with another and also with other climate forcing factors such as changes in the solar luminosity or atmospheric aerosol concentrations. Figure 10.6 displays direct radiative forcing estimates for various factors thought to be relevant to contemporary climate change, with error bars indicated. Carbon dioxide emissions are currently the single most important cause of positive radiative forcing at 1.6 W/m^2 (see Table 10.1). Taken together, the long-lived greenhouse gases are currently adding an additional 1 W/m^2 of energy into the troposphere compared with the unperturbed radiative balance. Anthropogenic effects on the surface albedo include decreases (cooling by increases in surface albedo) due to land use changes, and increases (warming by a darkening of the surface) due to the settling of black carbon on snow. The greatest anthropogenic cooling is that associated with aerosols. In Figure 10.6, it appears that the cooling effects of aerosols have the potential to "save the day," at least partially, from greenhouse gas–induced warming, and this is exactly why geoengineering techniques to ameliorate global warming often concentrate on manipulating atmospheric aerosol concentrations. However, recall that aerosol emission is often coupled with greenhouse gas emission. Also,

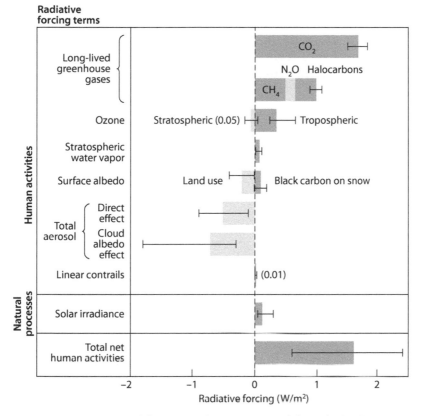

Figure 10.6. Summary of the principal components of the radiative forcing of climate change. Values represent the forcings in 2005 relative to the start of the industrial era (about 1750). From Climate Change 2007: The Physical Science Basis. Contribution of Working Group I to the Fourth Assessment Report of the Intergovernmental Panel on Climate Change (IPCC).

note the very large error bars in Figure 10.6 and recall that aerosol effects are regional. Despite the cooling effects of aerosols, in 2005 the net radiative forcing due to human activity was estimated to be 1.6 W/m².

10.3 EXERCISES

10.1. The increase in the longwave direct radiative forcing associated with an increase in atmospheric CO_2, $\Delta F(CO_2)$, is proportional to the natural log of the CO_2 concentration according to the formula

$$\Delta F(CO_2) = A \ln(C/C_0),$$

where $A = 6.3$ W/m², C_0 is the initial CO_2 concentration, and C is the final CO_2 concentration.

(a) Calculate the change in the direct radiative forcing, $\Delta F(CO_2)$, associated with CO_2 increases between 1800 and 2010.

(b) A reasonable prediction for the atmospheric CO_2 concentration for the year 2030 is 450 ppm. Calculate the increase in the radiative forcing associated with the predicted CO_2 increase between 2010 and 2030.

(c) Compare average time rates of change of $\Delta F(CO_2)$ for each of the two periods 1800–2010 and 2010–2030 in units of $W/(m^2 \cdot yr)$.

11

CLIMATE CHANGE PROCESSES

As discussed in the previous chapter and in chapter 4, the immediate effect of changes in atmospheric concentrations of greenhouse gases in the troposphere is an increase in longwave radiation—the greenhouse effect, or direct radiative forcing. Now, we consider the response of the climate system to that forcing.

11.1 CLIMATE SENSITIVITY

Defining a climate sensitivity parameter λ,

$$\lambda \equiv \frac{\partial T^*}{\partial F}, \tag{11.1}$$

allows us to compare and quantify the effects of various climate forcing factors on climate. In Eq. 11.1, F is a radiative forcing (W/m^2) and T^* is the globally and annually averaged surface air temperature. T^* is often used as a single-parameter representation of the climate state, though it could be replaced with any number of climate parameters, and F can represent various forcing factors. Equation 11.1 is an idealized representation of a very complex system, but it is useful for developing basic ideas of climate forcing and response.

As an example, consider the simplest representation of the climate system, namely, the global-mean radiative balance at the top of the atmosphere. Solving Eq. 4.6 for T_E provides an excruciatingly oversimplified model for the radiative equilibrium temperature:

$$T_E = \left[\frac{S_0(1-\alpha)}{4\sigma} \right]^{1/4}. \tag{11.2}$$

The sensitivity of T_E to changes in the solar constant is then

$$\lambda_{S_0} = \frac{\partial T_E}{\partial S_0} = \frac{S_0^{-3/4}}{4} \left[\frac{(1-\alpha)}{4\sigma} \right]^{1/4}. \tag{11.3}$$

According to Eq. 11.3, T_E is more sensitive to changes in the solar constant for smaller values of S_0 and larger values of α. With $\alpha = 0.31$ and $S_0 = 1368$ W/m^2,

$$\lambda_{S_0} = \frac{\partial T_E}{\partial S_0} = 0.046 \ K/(W \cdot m^2). \tag{11.4}$$

Keeping the same planetary albedo and reducing S_0 by 25% results in a stronger sensitivity:

$$\lambda_{S_0} = \frac{\partial T_E}{\partial S_0} = 0.058 \text{ K/(W} \cdot \text{m}^2). \tag{11.5}$$

The preceding calculation addresses the sensitivity of the entire climate system to changes in external solar forcing, but it is not helpful in evaluating the system's response to increasing greenhouse gas levels. The sensitivity of the surface climate to changes in greenhouse gas concentrations can be estimated using the equilibrium surface heat balance equation (Eq. 5.9, with Eq. 5.10) combined with a radiative model that accounts for changes in the longwave back radiation at the surface, F_{BACK}, due to increasing greenhouse gases. Given that the direct radiative effects of a doubling of tropospheric CO_2 is estimated to be an increase in F_{BACK} of 4 W/m² (see exercise 10.1), and assuming that all the other factors in the surface heat balance remain unchanged from the values given in Figure 5.4, the surface heat balance calculation gives the following results:
For today's climate,

$$T_S = \left[\frac{(1 - \alpha_S) S_{INC} + F_{BACK} - H_S - H_L}{\varepsilon \sigma} \right]^{1/4} = 289 \text{ K}. \tag{11.6}$$

For the doubled-CO_2 climate, increase F_{BACK} to 337 W/m² but hold all other values fixed. Then, $T_S = 290$ K.

According to this simple calculation, a doubling of the atmospheric CO_2 concentration is estimated to cause a 1 K warming of surface temperature. (This is in the absence of a climate system response, so stay tuned until the following section which deals with the issues of climate feedback processes.) By this rough estimate, the climate sensitivity to changes in the longwave back radiation, λ_{LW}, is

$$\lambda_{LW} \equiv \frac{1 \text{K}}{4 \text{ W/m}^2} = 0.25 \text{ K/(W/m}^2), \tag{11.7}$$

that is, for every 1 W/m² change in the longwave back radiation, T_S changes by 0.25 K.

How accurate are these calculations? Would you imagine, for example, that if the surface warmed even a little that the sensible and latent heat fluxes would remain the same? Of course not, as a quick reference to chapter 5 confirms. In fact, small changes in the direct radiative forcing trigger other changes in the climate system that greatly modify the climate's sensitivity to increasing greenhouse gases.

11.2 CLIMATE FEEDBACK PROCESSES

How does climate sensitivity come about? One way to organize your thoughts for investigating changes in the very complicated climate system is to study individual climate feedback processes in isolation. Several examples of climate feedbacks known to be important for determining climate are presented here. Each example assumes that the feedback is triggered by an increase in the atmospheric CO_2 (or other greenhouse gas) concentration and a small increase in the global-mean surface temperature, T^*, in response to the direct radiative forcing. But this kind of analysis can be applied to consider other climate variables and other forcing factors as well.

In the schematic drawings of *feedback loops*, climate factors are connected by arrows to indicate the order of the cause and effect. Each arrow is labeled with a + or a − sign, which indicates how the two factors are correlated. A + sign indicates that the factors are positively correlated, so that an increase in one leads to an increase in the other, or a decrease in one leads to a decrease in the other. A negative sign means that the factors are negatively correlated, so that an increase in the first factor leads to a decrease in the second factor, or vice versa.

WATER VAPOR–TEMPERATURE FEEDBACK

Figure 11.1a is a schematic representation of the chain of events in the *water vapor–temperature feedback*. Start at the top, and read this feedback loop as follows:

> An increase (decrease) in atmospheric CO_2 concentration leads to an increase (decrease) in T^*, which leads to an increase (decrease) in evaporation and atmospheric water vapor, which leads to an increase (decrease) in longwave back radiation at the surface, which leads to an increase (decrease) in T^*.

Recall that the positive and negative signs are not the sign of the change but, rather, the sign of the correlation.

The water vapor–temperature feedback is a *positive feedback*—the original T^* increase (or decrease) is amplified. This is a very powerful feedback process in the climate system. Estimates using climate models indicate that in the absence of this feedback, doubling atmospheric CO_2 concentrations would lead to an increase in T^* of approximately 1.2 K (similar to our simple calculation using the surface heat balance equation in the absence of climate feedbacks).

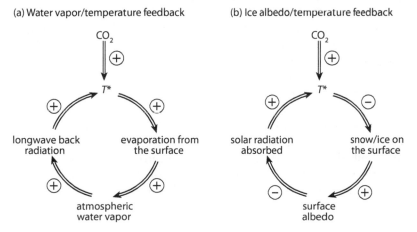

Figure 11.1 Schematic representations of the (a) water vapor/temperature and (b) ice albedo/temperature feedbacks.

But if surface evaporation is allowed to respond to the surface warming, and more water vapor is loaded into the atmosphere as a result (see Eq. 5.21), the warming is doubled or tripled.

Why doesn't this feedback continue indefinitely? Could this feedback run away, loading more and more water into the atmosphere and causing more warming of the planet? No. All other things being equal, the feedback must end when the air reaches saturation (see Fig. 2.32) and water condenses out as precipitation. But there are also competing processes that can contain water vapor levels below saturation, for example, the land surface may dry out or plants may resist evapotranspiration through *stomatal resistance*.

ICE ALBEDO–TEMPERATURE FEEDBACK

The *ice albedo–temperature feedback* is sketched in Figure 11.1b. Surface warming due to increases in atmospheric CO_2 melts permanent snow and ice, or delays the onset of the snow season, or hastens the end of winter. The resulting darkening of the surface (decrease in surface albedo) permits the absorption of more solar radiation, which further warms the surface. Conversely, decreases in T^* lead to more snow and ice on the surface, a higher albedo, and less solar absorption. This is another positive feedback, since it amplifies the initial temperature change.

Note the minus signs in the ice albedo–temperature feedback schematic. The first leg of the loop, marked with a minus sign, indicates that T^* and surface snow and ice change in opposite directions. An increase in T^* leads to a decrease in surface snow and ice, and a decrease in T^* leads to an increase in surface snow and ice. There is a second minus sign in the relationship between the surface albedo and the solar radiation absorbed, so the net effect is a positive feedback.

The ice albedo–temperature feedback introduces latitudinal dependence into the climate change signal, since this feedback does not operate at low latitudes, except near tropical mountain glaciers. This is one of the two processes responsible for the *polar amplification* of the global warming signal. The other is the vertical stability of the polar atmosphere, especially in winter. In Figure 2.9 we noted the presence of temperature inversions at high latitudes. When increased greenhouse gas concentrations increase the longwave back radiation to the surface, the associated warming is concentrated near the surface when vertical mixing is inhibited by these temperature inversions. The amplification of the global warming signal due to the ice albedo–temperature feedback tends to be strongest at lower latitudes, for example, near ice and snow margins, and larger than the amplification due to the polar inversion.

The ice albedo–temperature feedback also introduces seasonality into the climate change dynamics. For example, in middle latitudes this feedback amplifies the temperature response in the late fall and early spring, shortening the length of time that snow is on the land.

The ice albedo–temperature feedback changes the amount of solar radiation absorbed by the earth system by changing the planetary albedo. This modifies the radiative equilibrium temperature of the earth system.

CLOUD FEEDBACK PROCESSES

Because of the complexities in simulating clouds and evaluating their influence on climate, clouds are the greatest source of differences among climate model simulations of future climate, and their role in future climate change is uncertain. There are many different kinds of clouds with a wide range of albedos. Their distributions and the physical processes that create them vary with latitude and season.

Here, without much technical detail, is an overview of cloud feedback processes. These feedbacks are difficult to evaluate because clouds interact with both longwave and shortwave radiation. Clouds are important for determining the planetary shortwave albedo at all latitudes and, being composed of water vapor, ice, and liquid water, they also absorb and reemit longwave radiation. Clouds are also influenced by aerosol distributions and types, and they are coupled to the global water cycle.

As discussed in section 4.7, the global- and annual-mean cloud forcing (Eq. 4.38) is negative ($F_C = -17.3$ W/m^2), so the net influence of clouds is to cool climate. However, this does not imply that the cloud response to increasing greenhouse gases will provide a negative feedback, since F_C can increase (become less negative, or even become positive) or decrease as climate changes.

To study the role of clouds in climate change, various aspects of the cloud/climate interaction are considered individually. First, consider cloud amount. Will global cloud amounts increase or decrease in response to greenhouse gas forcing? The answer is not known and will depend on a consideration of various cloud types under various conditions. Assuming for the sake of argument (and in agreement with many, but not all, climate model simulations) that cloud amounts decrease when T^* increases in response to greenhouse gas forcing, we can diagram two feedback loops isolating cloud radiative processes—one for the consideration of longwave fluxes and one for shortwave fluxes—in Figure 11.2. Note that with this assumption changes in cloud amounts provide

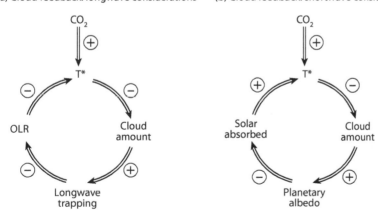

(a) Cloud feedback: longwave considerations   (b) Cloud feedback: shortwave considerations

Figure 11.2 (a) Longwave and (b) shortwave radiation feedback loops assuming cloud amount decreases with increasing temperature.

a negative feedback in their interaction with longwave radiation, and a positive feedback when shortwave processes are evaluated.

To increase realism, consider possible changes in cloud distributions, specifically, in cloud-top altitudes. If lapse rates remain constant, cloud-top temperatures decrease with increasing cloud top height (and vice versa). Because the optical depth of clouds is large (see chapter 4), we can assume that the longwave emission to space from clouds, as well as the reflection of solar radiation, occurs at cloud top. Thus, an increase in cloud-top altitudes makes the cloud a more effective longwave trapping agent (enhanced longwave effect), but it doesn't change how effectively the cloud cools climate (unchanged shortwave effect). A feedback loop is sketched in Figure 11.3a, assuming that cloud-top temperatures decrease as climate warms. This will be the case if a cloud layer is displaced upward (with no change in lapse rate) or if the atmosphere becomes more unstable in a way that deepens convection.

Another way to think about changes in cloud-top temperature in a global sense is to consider the possibility that low or high clouds may be preferentially modified as climate changes. For example, if greenhouse gas–induced climate change leads to a decrease in the coverage by low clouds with an increase in the coverage by high clouds, then the overall effect on climate will be a positive feedback.

Cloud albedo is another potential contributor to climate change. Here, we are concerned primarily with cloud microphysical properties, for example, the liquid water content of a cloud, droplet size distributions, and features of a cloud's ice particles. If the liquid water content of a cloud increases as climate warms, for example, its shortwave albedo will increase and provide a negative feedback on climate change (see Fig. 11.3b).

Changes in atmospheric aerosols may also feed back to climate in significant ways through their effects on clouds. This is the *indirect effect of aerosols* on climate, in contrast with the direct effects produced when aerosols absorb and scatter radiation. Natural and anthropogenic atmospheric aerosols serve as *cloud condensation nuclei* (CCN)—the basis for cloud seeding.

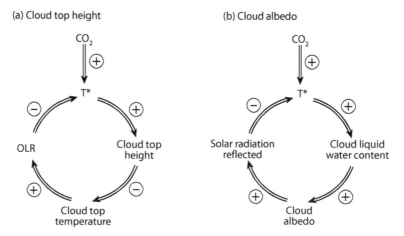

Figure 11.3 Example cloud feedback loops related to (a) cloud top height and (b) cloud brightness.

One example is a feedback involving the naturally occurring CCN dimethyl sulfide, or DMS, which is released into the atmosphere at the ocean surface by phytoplankton. When phytoplankton growth responds to changes in climate conditions, DMS emissions and cloud amounts or distributions could be effected. This feedback is the kind of process sometimes discussed in connection with the Gaia hypothesis (section 2.5).

The indirect effect of aerosols also includes changes in cloud albedo through modifications of cloud droplet size distributions, liquid water content, cloud height, and cloud lifetimes. The variety of aerosol types associated with human activity, as well as their complicated distributions in space and time, makes it difficult to evaluate these indirect effects on climate.

11.3 EXTREME HYDROLOGIC EVENTS

Many projections of future climate due to increases in greenhouse gases suggest that the global hydrologic cycle will intensify in the warmer climate and result in increased incidence of severe rainfall events, flooding, and drought. These projections are fairly robust in the sense that many different models under various assumptions about future emissions are in general agreement, but the projections may not apply regionally. Some models also predict that decreases in global- and annual-mean rainfall frequency will accompany the increases in the intensity of precipitation events.

Our physical understanding of how and why rainfall intensity may increase as climate warms derives from a scaling argument based on the Clausius-Clapeyron equation (Eq. 2.8). Recall that this equation expresses the relationship between the temperature of an isothermal, closed system consisting of moist air overlying a flat, quiescent water surface and the amount of water vapor in the overlying air. As the temperature of the system increases, evaporation increases, and more water vapor enters the air. When the rate of condensation onto the surface equals the evaporation rate, the saturated atmosphere will contain more water vapor at the new equilibrium.

Although the Clausius-Clapeyron equation is strictly applicable only to the closed system for which it is derived, it is often applied less formally to climate. For temperatures typical of the earth's surface, the saturation partial pressure of water vapor, e_s, increases exponentially with temperature (see Fig. 2.32)—about 7% for every 1 K warming. Because relative humidity is predicted to remain constant as climate warms, we can expect that the partial pressure of water vapor will increase exponentially, as will the specific humidity, q. According to Eq. 9.8, an increase in q leads to an increase in the column moisture convergence if the circulation remains unchanged. Of course, the circulation will not remain unchanged as the global climate warms, and this fact introduces uncertainty in projections of intensified precipitation and also suggests that changes in precipitation intensity will be regional.

Projections of increases in the occurrence of drought can be confusing because these events are often not precisely defined. There are four general types of drought, namely, meteorological, hydrological, agricultural, and socioeconomic. *Meteorological drought* refers to extended periods with negative precipitation anomalies, with a time scale dependent on regional climate statistics including local seasonality and interannual variability. *Hydrological drought* describes low levels in lakes, reservoirs, and the subsurface, and/or low flow rates in streams and rivers. *Agricultural drought* refers to low soil moisture values, or deficits in surface water availability, and may be defined relative to season and crop type. Socioeconomic drought is measured by its effects on people.

Attempts to quantify drought have led to the development of various drought indices such as the *Palmer Drought Severity Index (PDSI)*, which uses both precipitation and temperature information and is most useful for characterizing long-term drought. Another drought index, the *Standardized Precipitation Index (SPI)*, is thought to be more useful for identifying regions of emerging regional drought. It is based on calculations of the percentage of normal rainfall for a time scale and region defined by the user for a given application.

11.4 EXERCISES

11.1. Calculate the amount of longwave radiation leaving an area of 1 m² at the tropopause for three cases: cloudless atmosphere, low cloud, and high cloud. In each case, model the atmosphere as a transparent slab with a temperature structure similar to that observed for the troposphere. Choose reasonable temperatures for the surface, a low cloud, and a high (in the troposphere) cloud. Assume that the surface and clouds are perfect blackbodies.

11.2. Calculate the longwave radiative forcing due to an increase in the height of a midlatitude cloud from 4 to 4.5 km. Assume that the tropospheric lapse rate doesn't change. Refer to chapter 2 to choose reasonable values for the tropospheric lapse rate.

11.3. Daisyworld[1]

(a) The surface of Daisyworld I is covered by two species of daisies. One species is pure white, the other is pure black. The white daisies produce increasing numbers of seeds as the average surface air temperature rises above 15°C. The black daisies produce increasing numbers of seeds as the average surface air temperature falls below 15°C. For thousands of years the average surface air temperature of Daisyworld I was 15°C, and the populations of the two species of daisies were equal.

One day, a space traveler arrived to explore Daisyworld. She looked around a bit but, being allergic to daisies, quickly departed. The exhaust from her spaceship was highly concentrated CO_2. The CO_2 was transported throughout the atmosphere, and the average surface air temperature began increasing due to the direct radiative forcing. What happened on Daisyworld I?

(b) The surface of Daisyworld II is covered by two species of daisies. One species is pure white; the other is pure black. The white daisies

---

[1] Daisyworld is the creation of James Lovelock, author of the Gaia hypothesis.

produce increasing numbers of seeds as the average surface air temperature falls below 15°C. The black daisies produce increasing numbers of seeds as the average surface air temperature rises above 15°C. For thousands of years the average surface air temperature of Daisyworld II was 15°C, and the populations of the two species of daisies were equal.

One day, a space traveler arrived to explore Daisyworld. He looked around a bit but, being in a bit of a hurry, quickly departed. The exhaust from his space ship was highly concentrated CO_2. The CO_2 was transported throughout the atmosphere, and the average surface air temperature began increasing due to the direct radiative forcing. What happened on Daisyworld II?

12
CLIMATE SIMULATION AND PREDICTION

The feedback processes discussed in the previous chapter, along with many others, are triggered when a forcing is applied to the climate system. How can all these processes, and interactions among these processes, be accounted for to evaluate the resulting climate change? How do we know which processes are relatively important, either globally or regionally, and which are ineffective at changing climate? Computer climate models based on the equations and principles discussed in previous chapters are used to keep track of the feedbacks and interactions.

Climate models range from very simple representations of the earth system to very complex models that are run on the world's fastest computers. The most complete climate models are a comprehensive compilation of our knowledge of how the climate system works, cast in computer language to allow for numerical solution.

However complex, all climate models consist of a set of governing equations accompanied by a set of assumptions. We begin with the simplest climate model.

12.1 ZERO-DIMENSIONAL CLIMATE MODEL

The simplest climate model has zero dimensions—it treats the whole earth system as one point and is based on the concept of radiative equilibrium discussed in chapter 4. Two equations are solved simultaneously for two unknowns, the radiative equilibrium temperature (T_E) and the surface temperature (T_S). From section 4.2, the steady-state radiation balance at the top of the atmosphere is given by

$$\frac{(1-\alpha)S_0}{4} = \sigma T_E^4. \tag{12.1}$$

The second equation, which relates T_S and T_E, is generated from observations of today's climate, which indicate that the globally and annually averaged surface temperature of the earth is (roughly) 35 K warmer than the radiative equilibrium temperature:

$$T_S = T_E + 35 \text{ K}. \tag{12.2}$$

The set of assumptions that accompany the equations to define a climate model are important for defining the context of the model, including the bounds of its usefulness. For the simplest climate model, these assumptions are as follows:

(i) The earth system is in equilibrium with the sun, so there is no temperature trend and energy input = energy output.
(ii) Stefan's law holds, that is, the earth system can be treated as a blackbody (emissivity $\varepsilon = 1$).
(iii) The relationship between T_E and T_S is the same for various climate states.

The first assumption indicates that this climate model cannot provide information about transient climate states, that is, the evolution from one climate state to the next. It can provide information only about different equilibrium, or steady-state, climates.

Similar to the discussion of climate sensitivity in chapter 11, an analytical calculation using the simplest climate model is used to illustrate the basic methodology of climate model simulation. We use the model to simulate the consequences of a 10% increase in planetary albedo.

STEP 1. UNPERTURBED CLIMATE STATE

The first step in any climate model simulation is to produce a *control climate*, which is a representation of the unperturbed, or present day, climate. With present-day values for the albedo ($\alpha = 0.310$) and solar constant ($S_0 = 1368$ W/m^2), the simultaneous solution of Eqs. 12.1 and 12.2 gives $T_E = 254$ K and $T_S = 289$ K (or 16°C , or 61° F). This is the zero-dimensional model's control climate state.

STEP 2. MODEL EVALUATION

The second step is to evaluate the accuracy of the control simulation of the observed climate. For the zero-dimensional model, this involves comparing the observed globally and annually averaged T_E and T_S climatologies with those produced by the model. For this simple climate model, the circular nature of this validation is readily apparent, since one of the governing equations (Eq. 12.2) is empirically derived. Like most climate models, the simplest climate model is a blend of theoretical governing equations (Eq. 12.1) and empirical constants (Eq. 12.2). In more complex models, discussed below, the agreement with observations is not perfect and the influence of observed values is much weaker.

STEP 3. PERFORM THE EXPERIMENT

To address the question at hand—What are the consequences of a 10% increase in planetary albedo?—the value of α is increased by 10%, to 0.341. All other values are held fixed to isolate the effects of the albedo change. Solving Eqs. 12.1 and 12.2 again, but with the changed albedo, we get $T_E = 251$ K and $T_S = 286$ K. Thus, the answer to the question posed to this model is: A 10%

increase in the planetary albedo causes a cooling in the globally and annually averaged surface temperature of 3 K (3°C, or 6°F).

STEP 4. EVALUATE THE RESULTS

The final step, usually the most time consuming and interesting, is to evaluate the results. Do they make sense? How do they compare with results from other models and with our basic understanding of how climate works? What confidence can be ascribed to the results? Was the climate sensitive to the forcing? In this case, because a 10% increase in α caused a 3 K cooling of global climate (as represented by the surface temperature), we would say that climate is sensitive to changes in the planetary albedo of even a few percent.

12.2 SURFACE HEAT BALANCE CLIMATE MODELS

The zero-dimensional climate model addresses the relationship between the globally and annually averaged surface temperature and the average albedo, and between surface temperature and the solar constant. But this model is useless for studying the influence of increases in greenhouse gases on climate. However, we can construct another analytical model based on the surface heat balance discussed in chapter 5 to study the response to the greenhouse effect, or changes in the longwave back radiation due to increasing greenhouse gas concentrations.

We used this model in section 11.1 to explore the climate sensitivity to direct radiative forcing. Many unrealistic assumptions were needed to use Eq. 5.9 as a climate model, including that α, H_S, and H_L remain constant despite the enhanced longwave radiative heating of the surface. We used this framework to estimate the surface temperature response to direct radiative forcing in the absence of climate feedbacks, so the calculation is seriously flawed as a climate predictor. None of the important climate feedback mechanisms discussed in chapter 11 were allowed to operate since none of the other surface heat flux terms changed.

To refine this calculation, and better predict the system response, it is necessary to take climate feedback processes into account. For example, the turbulent heat fluxes, H_S and H_L, depend strongly on T_S (section 5.3); and F_{BACK} will change according to the water vapor–temperature feedback mechanism when evaporation rates and, therefore, atmospheric water vapor rates increase due to surface warming. When surface temperatures change the ice albedo feedback will also operate, affecting the value of α. There also will be complex changes in the properties and distribution of the world's clouds. Changes in surface winds and in the hydrologic cycle will force changes in the ocean circulation and in land surface conditions as well. These and many other processes must be accounted for in a more accurate and complete projection of future climate.

12.3   GENERAL CIRCULATION MODELS

To account for feedback processes, nonlinearities, and climate subsystem interactions, computer models of the global climate known as *general circulation models*, or GCMs, are constructed. GCMs are mathematical compilations of our understanding about how the climate system works, and they are our most sophisticated tool for predicting future climate.

Given a set of governing equations, accompanied by physical parameterizations for processes that cannot be explicitly calculated, GCMs produce solutions for the climate state that are consistent with a set of climate forcing factors. Climate prediction is fundamentally different from weather prediction, which uses computer models initialized with the observed state of the atmosphere to advance the initial condition through time to produce predictions for a few days in the future. Climate simulations, in contrast, should not be highly dependent on the initial state.

GCMs can be categorized according to how many and which subsystems of the full climate system (see chapter 1) are included.

ATMOSPHERIC GCMS

The governing equations of an *atmospheric GCM*, or AGCM, are listed in Box 12.1. They may be written in a different form, for example, using different coordinate systems, for different models, but the underlying physics is the same.

Box 12.1   Governing Equations of an Atmospheric GCM

$$\frac{\partial u}{\partial t} = -\vec{v} \cdot \nabla u + \frac{uv\tan\phi}{a} - \frac{uw}{a} - \frac{1}{\rho a \cos\phi}\frac{\partial p}{\partial \lambda} + fv - 2\Omega w\cos\phi + F_\lambda. \quad (12.5)$$

$$\frac{\partial v}{\partial t} = -\vec{v} \cdot \nabla v - \frac{u^2\tan\phi}{a} - \frac{vw}{a} - \frac{1}{\rho a}\frac{\partial p}{\partial \phi} - fu + F_\phi. \quad (12.6)$$

$$\frac{\partial w}{\partial t} = -\vec{v} \cdot \nabla w + \frac{u^2 + v^2}{a} - \frac{1}{\rho}\frac{\partial p}{\partial z} - g + 2\Omega u\cos\phi + F_z. \quad (12.7)$$

$$\frac{\partial T}{\partial t} = -\vec{v} \cdot \nabla T + \frac{\omega}{\rho c_p} + \frac{Q}{c_p}. \quad (12.8)$$

$$\frac{1}{\rho}\frac{d\rho}{dt} + \nabla \cdot \vec{v} = 0. \quad (12.9)$$

$$p = \rho RT. \quad (12.10)$$

$$\frac{\partial(\rho q)}{\partial t} = -\vec{v} \cdot \nabla(\rho q) + E - P. \quad (12.11)$$

- Eqs. 12.5–12.7 are Newton's second law ($F = ma$) in three dimensions as applied to a fluid (i.e., including the advection terms), as derived in chapter 6.
- Eq. 12.8 is the first law of thermodynamics (Eq. 5.3).
- Eq. 12.9 is the *continuity equation*, an expression of conservation of mass.
- Eq. 12.10 is the ideal gas law (Eq. 5.1).
- Eq. 12.11 is an equation for the conservation of water mass for a parcel of air (see chapter 9).

The model equations do not include explicit calculations for friction (F^λ, F^φ, and F^z), the diabatic heating rate (Q), precipitation (P), and evaporation (E); these terms are represented by physical parameterizations. Parameterizations for friction are based on basic principles of momentum diffusion, which occur on space scales much smaller than those resolved by GCMs (discussed below). Convection may also occur on spaces scales smaller than resolved and when this is the case—as it generally is—parameterizations are used to calculate convective vertical velocities, water condensation rates, and the resulting formation of clouds. The parameterization of convection and condensational heating also yields the precipitation rate. Evaporation rates are provided by coupling the AGCM with a land surface model, as discussed below.

The AGCM's governing equations (Eqs. 12.5–12.11) are solved simultaneously at grid points for the *tendencies* (Eulerian time rates of change), which are added to the most recent solution to advance the climate state in time. Typical horizontal grid spacings are 100–200 km with 30–50 vertical levels. Typical time steps are about 10 minutes.

AGCMs use fixed sea surface temperatures from observations to specify the lower boundary condition over the ocean, or a relatively simple heat balance equation (e.g., Eq. 5.22) to represent a mixed-layer, or slab, ocean with heat capacity but no circulation. A more complete treatment of the climate system requires coupling the AGCM with its ocean counterpart.

OCEAN GCMS

Ocean GCMs (OGCMs) are governed by a set of equations that are very similar to those used in AGCMs (Box 12.1) but with some important differences. Because the ocean cannot be treated as an ideal gas (chapter 5), Eq. 12.10 is replaced by an equation of state for the ocean that includes the dependence of water density on salinity in addition to temperature and pressure. Also, Eq. 12.11, for the conservation of water vapor, is replaced by an equation for the conservation of salt,

$$\frac{\partial(\rho S)}{\partial t} = -\vec{u} \cdot \nabla(\rho S) - f(P) + f(E) - f(R), \tag{12.12}$$

where S is salinity and the functions $f(P)$, $f(E)$, and $f(R)$ represent the influence of precipitation, evaporation, and runoff (as from rivers, for example), respectively, on the salinity distribution.

LAND SURFACE MODELS

Land surface models (LSMs) are coupled to AGCMs to provide a more complete treatment of land surface/atmosphere interactions. These models have become increasingly sophisticated in recent years, and they replace the simple surface heat balance equation (Eq. 5.22) and so-called *bucket model* for surface hydrology. In the bucket model, a unit area of land is visualized as a bucket with a certain depth, for example, 15 cm, that represents the water-holding capacity of soil. As the GCM is integrated over time, the bucket model keeps track of the net delivery of water to the surface (precipitation minus evaporation). If that value exceeds the depth of the bucket, the water is assumed to run off to the sea.

Contemporary LSMs, such as the model sketched in Figure 12.1a, govern the exchanges of water (precipitation, evapotranspiration, and sublimation) and energy (latent and sensible heat fluxes, interactions with longwave and shortwave radiation) between the atmosphere and the land surface. Momentum exchanges through the treatment of surface friction are also included in detail. The surface may be covered with natural vegetation, managed vegetation, or buildings and cement as in an urban environment.

In the bucket model, no accounting is made of river routing or the possible reabsorption of runoff by surrounding land regions—the water is simply deposited into the ocean. The more complex LSMs represent the land surface with multiple soil layers through which heat diffuses and water infiltrates (Fig. 12.1b). Surface runoff is accounted for and may be organized into streams and rivers to allow for reabsorption of the water before it reaches the ocean, and flooding is simulated. The vegetation type is calculated as a function of the modeled climate state in *dynamic vegetation models* (DVMs).

ICE MODELS

The formation of ice and snow at the surface is important for simulating climate at high latitudes and high altitudes. A calculation for snow and glaciers is included in LSMs, and a treatment of sea ice is coupled to OGCMs.

Information about atmospheric temperature and precipitation can be combined to simulate snowfall rates within the AGCM and pass them into the LSM. The surface temperature calculation is used to determine a rate of snowmelt, which adds to surface moisture, and snow sitting on the surface is "aged" in the sense that its initially bright albedo of about 0.8 is decreased to an old-snow albedo of 0.6.

Land-based glaciers are another important component of the cryosphere (see section 2.4). These slowly varying components of the climate system may be prescribed in shorter simulations, or their observed distributions may be prescribed as initial conditions since snowfall over many thousands of years is needed in some regions (e.g., Antarctica) to build glaciers. Glacier change may be parameterized as a function of surface and atmospheric temperatures and

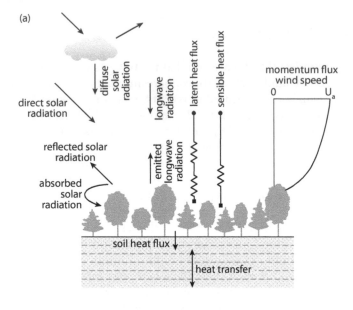

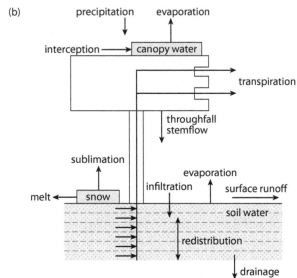

Figure 12.1 Schematic representations of (a) energy, moisture, and momentum transfers between the atmosphere and land surface and (b) the exchange of water between the atmosphere and the land surface. With permission from Gordon Bonan.

low-level wind and specific humidity. More complex glacier models take flow into account in addition to thermodynamics in calculating a glacier's mass balance. Dynamic glacier models must be integrated at much higher resolutions—a few tens of meters is typical.

There is intense interest in sea ice models because of the significant role of ice in determining climate and also because of the importance of predicting future sea ice distributions in association with the polar amplification of the global warming signal.

The formation of sea ice is very different from the formation of ice on land. The most complete sea ice models are based on a set of differential equations. Similar to the equations that govern AGCMs and OGCMs (Box 12.1), two of the governing equations for a sea ice model are the thermodynamic equation based on the first law of thermodynamics and a continuity equation expressing conservation of mass. However, the momentum equations that govern the movement of sea ice are highly nonlinear and non-Newtonian. They are taken from the equations that govern plastic flow, which describes the complicated ice interactions that occur in nature.

The AGCM and the sea ice model interact not only through the surface temperature field but also through the wind field, which modifies the ice dynamics. The sea ice model must also be coupled both dynamically and thermodynamically to the OGCM. Winds and currents partly determine how ice breaks up and the degree to which collisions occur in the sea ice field. In addition, brine exclusion processes, which are temperature dependent, modify the salinity of the ocean waters and influence the thermohaline circulation so the microphysics of ice formation must be included.

12.4 REGIONAL CLIMATE MODELS

Climate modeling at higher resolution than is typical for global models is useful for more accurately representing climate processes and for producing information about climate change on space scales that are more relevant for impacts analysis and policy decisions. For example, a climate model with grid spacing of 4 km or less can resolve convective processes within the governing equations (Box 12.1) and can become independent of the convective parameterizations known to be responsible for significant model error. Computational resources currently limit this level of resolution at the global scale; it is used in only a few high-resolution global simulations. But the governing equations can be applied over a regional domain and achieve this resolution. The trade-offs are loss of global connectivity and the need to make assumptions about conditions on the lateral boundaries of the domain.

12.5 EARTH SYSTEM MODELS

GCMs represent the physical climate system, but they do not track the flow of carbon and other elements among the climate system components. The atmospheric concentrations of carbon dioxide, methane, and other greenhouse gases are specified, often as a function of time, in GCM simulations.

To portray the climate system more completely, including the flow of chemical elements, *earth system models* (ESMs) combine fully coupled GCMs (including AGCMs, OGCMs, LSMs, and ice models) with *biogeochemistry* models that track the cycling of chemical elements in the climate system. Figure 1.2 includes many of the processes determined by biogeochemical cycling, which are indicated by flux arrows labeled with compounds such as CO_2, N_2O, and CH_4.

THE CARBON CYCLE

To simulate atmospheric CO_2 concentrations interactively in a model, the carbon cycle must be included. Figure 12.2 is a schematic overview of the global carbon cycle with estimated reservoirs and fluxes of carbon.

By far, the greatest reservoir of carbon on the planet is the earth's *lithosphere*, which consists of the crust and the upper mantle, where carbon constitutes 0.032% of the mass. The world's oceans are estimated to contain 38,000 Pg of carbon (PgC, with 1500 PgC sequestered in soils and 560 PgC in plants. About 4000 PgC remains in unused fossil fuels, in the form of coal, petroleum (crude oil, oil shale, oil sands), and natural gas. The atmosphere currently contains approximately 830 PgC.[1]

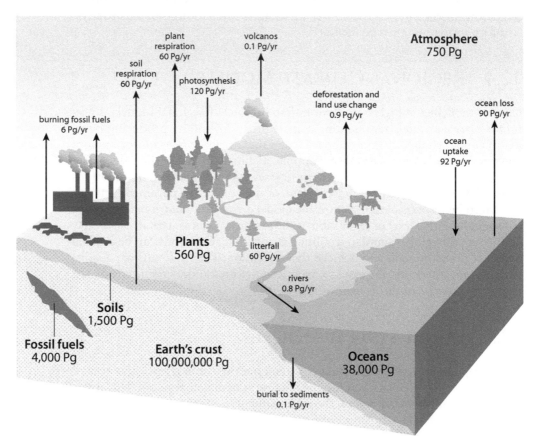

Figure 12.2 Estimated reservoirs and fluxes of the global carbon cycle. (NASA GLOBE program: www.nasa.gov.)

---

[1] The conversion from units of PgC to an atmospheric CO_2 concentration in ppm as shown in Fig. 10.1 is as follows: 1 ppm CO_2 = 7.822 Pg of CO_2 = 2.134 Pg of C.

Plant and soil respiration each release about 60 PgC to the atmosphere each year, and this source is balanced by the carbon capture of 120 PgC/yr during photosynthesis. Deforestation and land use are disturbing this balance and causing the additional release of an estimated 0.9 PgC/yr into the atmosphere.

Approximately 90 PgC/yr is released from the ocean to the atmosphere, and the ocean uptake is slightly larger, at 92 PgC/yr. The largest single source of atmospheric carbon (i.e., CO_2) is outgassing from the ocean. Note that the exchange of carbon between the atmosphere and the ocean is unbalanced. The oceans are currently estimated to be absorbing an extra 2 PgC from the atmosphere each year (see section 2.5 on the ocean biological pump). Human activity is perturbing the natural carbon cycle, leading to the imbalance in the atmosphere/ocean exchange by the release of 6 PgC/yr in fossil fuel burning and 0.9 Pg/C/yr in modifying the land surface and vegetation.

Figure 12.1 is highly simplified, and there are many complex processes involved as carbon in various forms is sequestered and flows among the climate subsystems. The partitioning of CO_2 between the atmosphere and the ocean, for example, is not only temperature dependent but also affected by the great variety of inorganic reactions and biological activity in the upper ocean and rates of exchange with the thermocline and deep ocean. The current challenges for including the carbon cycle in climate models are to identify the dominant processes that determine the sizes of carbon reservoirs and their rates of exchange, and to express them as accurately as possible in models.

THE METHANE CYCLE

The methane (CH_4) cycle is also included in ESMs because of methane's high GWP (Table 10.1), but our current knowledge of global methane cycling is incomplete. For example, the reason for the hiatus in global CH_4 trends from 2000 to 2006 seen in Figure 10.2 is not completely understood.

Estimates for components of the global CH_4 budget are listed in Table 12.1. Methane emissions derive from both biogenic and non-biogenic processes. The primary natural source of atmospheric methane, estimated at about 79% of current natural emissions, is wetlands. Other important natural sources include termites—CH_4 is produced when cellulose is broken down in their digestive tract—forests, and oceans.

Natural gas is composed of about 75% methane, so its use in human activity contributes to the observed increases in atmospheric CH_4. The largest atmospheric CH_4 source associated with the use of natural gas is fugitive emissions, including leaks from equipment and underground pipes. Vented emissions in natural gas recovery and combusted emissions also contribute significantly. Anthropogenic emissions also include rice agriculture, biomass burning, ranching (CH_4 is released as a result of enteric fermentation, which is the digestive decomposition of carbohydrates), and landfills and waste treatment facilities where bacteria release methane as they feed on organic material.

The primary sink for atmospheric CH_4 is its chemical reaction with the hydroxyl radical (OH) to form two other greenhouse gases, namely, water vapor and CO_2. Because OH is a scavenger species that reacts with and removes pollutants, increases in atmospheric CH_4 can amplify pollution levels by reducing

Table 12.1. Some estimated global methane sources and sinks

Natural sources (TgCH$_4$/yr)		Anthropogenic sources (TgCH$_4$/yr)		Sinks (TgCH$_4$/yr)	
Wetlands	174	Fossil fuels	172	Tropospheric OH	467
Termites	22	Rice agriculture	54	Stratosphere	39
Ocean	10	Biomass burning	47	Soils	30
Geologic	9	Ranching/ruminants	84		
Forests	5	Landfills/waste	54		
Total	220	Total	411	Total	536

Source: Based on Table 7.6 in IPCC, 2007.

OH concentrations. (The OH radical is very difficult to observe because it reacts quickly in the atmosphere. Its levels are inferred by comparing emission levels and atmospheric levels of gases for which reactions with OH are the primary sink, such as methyl chloroform. A trend in OH has not been identified.) Other sinks of atmospheric CH$_4$ are oxidation in soils and migration into the stratosphere.

THE NITROGEN CYCLE

For the climate change problem, global cycling of nitrogen (N) is modeled to understand how the concentration of atmospheric N$_2$O is determined. The nitrogen cycle must also be studied to understand the formation of NO$_x$ compounds for pollution studies. The nitrogen cycle consists of myriad paths and processes as nitrogen in various compounds passes among plants, animals, soils, atmosphere, and water. A brief overview is provided here.

The atmosphere is 78% molecular nitrogen (N$_2$), which is largely inert. This molecular nitrogen enters the nitrogen cycle when it is deposited into the soil by rain and, to a lesser extent, in dry deposition. Legumes (e.g., peas, beans, alfalfa, clovers, vetch), blue-green algae, and lichens convert molecular nitrogen into forms that plants can use, primarily ammonia (NH$_3$), to produce proteins and amino acids. This *nitrogen fixation* is accomplished by bacteria, for example, those that live in legume root nodules or freely in the soil. Other nitrifying bacteria in the soil convert ammonia into nitrites and nitrates, forms of nitrogen that plants can use to produce proteins and amino acids. The nitrogen cycles through animals when they eat plants, and their residues (excrement and bodies) are decomposed by bacteria and fungi. Nitrogen is lost from this soil-based cycle in *denitrification*, the process by which bacteria convert nitrogen to nitrous oxide (N$_2$O) and others forms of nitrogen for emission to the atmosphere; in *ammonia volatilization*, which is essentially the evaporation of ammonia gas from urea on the surface; and in leaching, when nitrates are washed below the root zone.

A similar nitrogen cycle occurs in the ocean. Soils covered by natural vegetation are estimated to produce 6.6 TgN (teragrams, or 10^{12}g, of nitrogen) each year, and the oceans add approximately 3.8 TgN/yr. These surface emissions

Table 12.2. Estimates of global N_2O sources

Natural sources	(TgN/yr)
Soil	6.6
Oceans	3.8
Atmospheric chemistry/lightning	0.6
Total	11.0

Anthropogenic sources	(TgN/yr)
Fossil fuels	0.7
Agriculture	2.8
Biomass burning	0.7
Human waste	0.2
Rivers, estuaries, coastal waters	1.7
Atmospheric deposition	0.6
Total	6.7

Note: Based on Table 7.7 in IPCC, 2007.

account for about 95% of the natural production of N_2O; the remaining 5% is produced by chemical reactions in the atmosphere (see Table 12.2).

Natural nitrogen fixation in soils is insufficient to support modern agricultural practice, so nitrogen fertilizer is added to agricultural soils. This addition enhances the natural nitrogen cycle in soils and, consequently, N_2O emissions (Table 12.2). In addition, agricultural activity amplifies the natural source of N_2O from rivers, estuaries, and the coastal ocean when runoff from agricultural land increases the nitrogen loading. Fossil fuel burning also produces atmospheric N_2O as a combustion product. Municipal waste, in the form of sludge from waste management facilities, is another anthropogenic source of N_2O.

As one can easily deduce from the numbers in Table 12.2, human activity has significantly perturbed the global nitrogen cycle, and this is why atmospheric N_2O concentrations are increasing (see Fig. 10.4).

This overview provides only the briefest introduction to the biogeochemical cycles of radiatively active components of the earth's atmosphere. The processes are complex, and developing accurate and complete models will require not only an accounting of the various sources and sinks but knowledge of the nonlinear organic and inorganic interactions among natural and anthropogenic systems. Global biogeochemical modeling is in the early stages of development. It is an important area of research needed to complete our understanding of the climate system and refine projections of the future.

12.6   EVALUATING MODEL UNCERTAINTY

Although climate models are multifaceted and sophisticated, the complexity of the climate system may seem overwhelming, as Figure 1.2 reminds us. Clearly, however, some processes are more important than others. Rather than attempting to include all the details of climate system interactions, which may be all but impossible, the thoughtful study of climate and climate change seeks to identify the most important processes and to include them accurately in models.

Projections of future climate are most useful for impacts analysis and decision making when they are accompanied by an evaluation of uncertainty. Comparisons of historical model runs with observations lend confidence to the ability of a given model to predict future climate. However, an accurate simulation of today's climate does not guarantee an accurate prediction of the future because different forcing factors may be dominant in the future and some empirically derived factors in model parameterizations may not be applicable to future climate.

Model confidence is also developed through the use of *ensemble* simulation techniques. A simulation may be rerun several times with initial conditions adjusted slightly to introduce noise, or with different versions of physical parameterizations to test the model's sensitivity. The model projections from each ensemble member can then be averaged to produce a prediction, and the spread among the ensemble members provides a measure of uncertainty.

Another way of building confidence in climate model predictions is to compare many different models. Although GCMs are all governed by the same basic equations (Box 12.1), each model employs different numerical integration schemes, smoothing and, especially, physical parameterizations. In fact, the most important reason for disparity among AGCM simulations has been attributed to the parameterization of clouds and convection. If the model predictions agree with one another despite the differences in parameterizations and other factors, it suggests that these "discretionary" model features are not dominating the projected climate change.

12.7 REFERENCE AND ADDITIONAL READING

IPPC, 2007: Contribution of Working Group I to the Fourth Assessment Report of the Intergovernmental Panel on Climate Change. Solomon, S., D. Qin, M. Manning, Z. Chen, M. Marquis, K. B. Averyt, M. Tignor, and H. L. Miller (eds.). Cambridge University Press, Cambridge and New York.

12.8 EXERCISES

12.1. Consider the zero-dimensional climate model.

(a) Derive an expression for the climate sensitivity parameter

$$\lambda_\alpha \equiv \frac{\partial T_E}{\partial \alpha}$$

that quantifies the sensitivity of T_E to α.

(b) Answer the following questions based on the results from (a): What is the sign of λ_α, and what does the sign imply about the relationship between changes in T_E and α?

Does the climate sensitivity to α depend on solar luminosity? If yes, in what way?

Does the climate sensitivity to α depend on α itself? If yes, in what way?

(c) Calculate the change in surface temperature due to a change in the planetary albedo from 0.31(today's value) to 0.25, according to this simple climate model.

(d) According to this climate model, what percent change in the solar constant would be required to cause an increase of 3 K in the surface air temperature?

(e) Would the climate model discussed in this problem be useful for understanding how the observed changes in greenhouse gas concentrations in the atmosphere may change climate? Why or why not?

12.2. Two insulated coolers with their tops off have been left in the middle of the Sahara Desert all day. The coolers measure 0.5 m × 1 m on top, and they are 0.5 m high. One cooler is filled with dry sand, and the other is filled with water. The content of each cooler has reached an isothermal temperature of 300 K.

(a) Use a time-dependent surface energy balance model to estimate how long will it take the contents of each insulated cooler to reach a temperature of 290 K after the sun sets. (Make assumptions as needed, and use the figures of chapter 2.)

(b) Which terms in the surface heat balance are the most important for this case?

(c) What is the poorest assumption you made?

(d) Relate your results to the concept of "continentality."

(e) How would the problem be different if the coolers were located in Chicago instead of the Sahara Desert?

UNITS, CONSTANTS, AND CONVERSIONS

UNITS:

Temperature:
   Celsius (°C) = Kelvin (K) − 273.15
   °C = (5/9) (°F − 32)
   temperature interval °C = °K = 1.8°F
Distance: meter (m) or kilometer (km)
   1 mile = 1.6 km
Wind speed: meters per second (m/s)
   1 mi/hr = 0.446 m/s
Angular velocity: radians per second (s^{-1})
Precipitation rate: mm/day or cm/day
Pressure: 1 millibar (mb) = 10^2 Pa (pascals) = 10^2 kg/(m · s^2) = 1 hPa (hectapascal)
Energy: joules (J) = (kg · m^2)/s^2
Heating rates: joules per second (J/s) = Watts (W)
Radiation flux density: heating rate per unit area [W/m^2 = J/(m^2 · s)]
Atmospheric concentration: parts per million by volume (ppm)
      parts per billion by volume (ppb)
      parts per trillion by volume (ppt)
Salinity: practical salinity units (psu)

CONSTANTS, WITH THE SYMBOLS USED IN THIS BOOK:

Radiation constants:
   Stefan-Boltzman constant (σ) = 5.67×10^{-8} W/(m^2 · K^4)
   Planck's constant (h) = 6.626×10^{-34} J · s
   Boltzmann constant (k) = 1.38×10^{-23} J/K
   speed of light (c) = 3.00×10^8 m/s
The Sun:
   solar "constant" (S_0) = 1368 W/m^2
   solar luminosity (L_S) = 3.9×10^{26} J/s
   solar radius = 6.96×10^5 km
   average Earth-sun distance = 1.50×10^8 km

The Climate System:

angular velocity of the earth (Ω) = 2π radians/24 hours = 7.3×10^{-5} s^{-1}

acceleration due to gravity (g) = 9.81 m/s^2

average radius of the earth (a) = 6371 km (equatorial radius is 6378 km)

latent heat of fusion (L) of water at 0°C = 3.34×10^5 J/kg

gas constant

$\quad R_d$ = dry air: 287 J/(kg · K)

$\quad R_v$ = water vapor: 461 J/(kg · K)

latent heat of vaporization (L) of water

°C	K	L (10^6 J/kg)
0	273.2	2.501
10	283.2	2.476
20	293.2	2.453
30	303.2	2.432

specific heat capacity (c)

specific heat at constant pressure for air (c_p) = 1004 J/(kg · K)

specific heat at constant volume for air (c_p) = 717 J/(kg · K)

water = 4218 J/(kg · K)

dry sand = 840 J/(kg · K)

density (0°C, 1000 hPa)

air = 1.275 kg m^{-3}

water = 1.000×10^3 kg m^{-3}

sand = 2.65×10^3 kg m^{-3}

COORDINATE SYSTEMS

Local Cartesian and Earth-centered spherical coordinate systems are defined for expressing equations within the earth system. These are not the standard Cartesian and spherical coordinate systems found in math text books. Rather, they are designed to easily relate to latitude and longitude coordinates.

LOCAL CARTESIAN COORDINATES

When space scales are relatively small, so the curvature of the Earth can be neglected, we can use the local Cartesian coordinate system shown in Figure B.1.

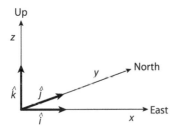

Figure B.1. The local Cartesian coordinate system. Orthogonal coordinate axes are x, y, and z, with unit vectors $\hat{i}$, $\hat{j}$, and $\hat{k}$, respectively.

- The east–west velocity, $u \equiv dx/dt$, is called the *zonal wind* in the atmosphere and the *zonal current* in the ocean. It is positive when it is directed in the positive x-direction, which is eastward (in the direction of the rotation of the earth). An eastward (westward) flow is also referred to as *westerly* (*easterly*) flow.
- The north/south velocity, $v \equiv dy/dt$, is called the *meridional wind* or the *meridional current*. It is positive when it is directed in the positive y-direction, which is northward. A northward (southward) flow is also referred to as *southerly (northerly)* flow.
- The *vertical velocity*, $w \equiv dz/dt$, is positive when it is directed in the positive z-direction, which is upward. Downward flow is sometimes referred to as *subsidence*, especially when sinking motion occurs on large space scales.

As discussed in chapter 2, z can be replaced by p as a vertical coordinate, in which case the vertical p-velocity, $\omega \equiv dp/dt$, is used to express vertical motion, and the x- and y-directions are parallel (orthogonal) to pressure surfaces instead of elevation surfaces. In this case, $\hat{i}$, $\hat{j}$, and $\hat{k}$ are still used as the unit vectors.

Vector Operations in Local Cartesian Coordinates:

Del operator: $\nabla \equiv \hat{i}\dfrac{\partial}{\partial x} + \hat{j}\dfrac{\partial}{\partial y} + \hat{k}\dfrac{\partial}{\partial z}$ $\qquad\qquad$ (B.1)

Horizontal del operator: $\nabla_h \equiv \hat{i}\dfrac{\partial}{\partial x} + \hat{j}\dfrac{\partial}{\partial y}$

∇_h is also written ∇_z, where the subscript z indicates that z is held constant or, when pressure is used as a vertical coordinate, ∇_p.

When the del operator acts on a scalar, s, the result is the gradient:

$$\nabla s \equiv \hat{i}\frac{\partial s}{\partial x} + \hat{j}\frac{\partial s}{\partial y} + \hat{k}\frac{\partial s}{\partial z}. \tag{B.2}$$

There are two ways to multiply two vectors, $\vec{A} \equiv \hat{i}A_x + \hat{j}A_y + \hat{k}A_z$ and $\vec{B} \equiv \hat{i}B_x + \hat{j}B_y + \hat{k}B_z$. The cross-product is defined as

$$\vec{A} \times \vec{B} = \begin{vmatrix} \hat{i} & \hat{j} & \hat{k} \\ A_x & A_y & A_z \\ B_x & B_y & B_z \end{vmatrix} = \hat{i}(A_yB_z - A_zB_y) + \hat{j}(A_zB_x - A_xB_z) + \hat{k}(A_xB_y - A_yB_x), \tag{B.3}$$

and the result is a vector. Equivalently,

$$\vec{A} \times \vec{B} = |\vec{A}||\vec{B}|\sin\theta, \tag{B.4}$$

where $|\vec{A}| \equiv \sqrt{A_x^2 + A_y^2 + A_z^2}$ and θ is the angle between $\vec{A}$ and $\vec{B}$. The cross product picks up the components of $\vec{A}$ and $\vec{B}$ that are perpendicular to each other.

The cross product is also used to calculate the "*curl*":

$$\nabla \times \vec{A} = \begin{vmatrix} \hat{i} & \hat{j} & \hat{k} \\ \dfrac{\partial}{\partial x} & \dfrac{\partial}{\partial y} & \dfrac{\partial}{\partial z} \\ A_x & A_y & A_z \end{vmatrix} = \hat{i}\left(\frac{\partial A_z}{\partial y} - \frac{\partial A_y}{\partial z}\right) + \hat{j}\left(\frac{\partial A_x}{\partial z} - \frac{\partial A_z}{\partial x}\right) + \hat{k}\left(\frac{\partial A_y}{\partial x} - \frac{\partial A_x}{\partial y}\right) \tag{B.5}$$

The second way to multiply two vectors is to use the dot product:

$$\vec{A} \cdot \vec{B} = A_xB_x + A_yB_y + A_zB_z. \tag{B.6}$$

Note that the result of the dot product is a scalar. Equivalently,

$$\vec{A} \cdot \vec{B} = |\vec{A}||\vec{B}|\cos\theta. \tag{B.7}$$

The dot product picks up the components of $\vec{A}$ and $\vec{B}$ that are parallel to each other. The dot product is used to calculate the divergence

$$\nabla \cdot \vec{v} = \frac{\partial u}{\partial x} + \frac{\partial v}{\partial y} + \frac{\partial w}{\partial z}. \tag{B.8}$$

Combinations are also possible, for example,

$$\hat{k} \cdot (\nabla \times \vec{v}) = \begin{vmatrix} 0 & 0 & \hat{k} \\ \dfrac{\partial}{\partial x} & \dfrac{\partial}{\partial y} & \dfrac{\partial}{\partial z} \\ A_x & A_y & A_z \end{vmatrix} = \frac{\partial v}{\partial x} - \frac{\partial u}{\partial y} \tag{B.9}$$

where $\vec{v} \equiv u\hat{i} + v\hat{j} + w\hat{k}$. In this case, the dot product is used to find the vertical component of the curl of the velocity. This quantity is used frequently in earth system dynamics and is known as the relative vorticity.

EARTH-CENTERED SPHERICAL COORDINATES

For large-space scales, the curvature of the earth must be taken into account, so the spherical coordinate system shown in Figure B.2 is used. To locate point

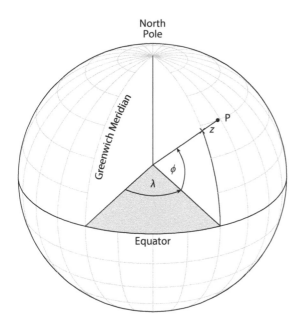

Figure B.2. The Earth-centered spherical coordinate system.

P in this coordinate system, latitude (ϕ) and longitude (λ) are defined for the point (marked by an x) on the surface below the point (or above it for an ocean application), that is, making the thin atmosphere (ocean) approximation. The vertical coordinate, z, is the elevation of the point above the earth's surface.

Note that it is easy to translate between the horizontal components of the local Cartesian and the earth-centered spherical coordinates. With the thin atmosphere approximation, Figure B.3 indicates that

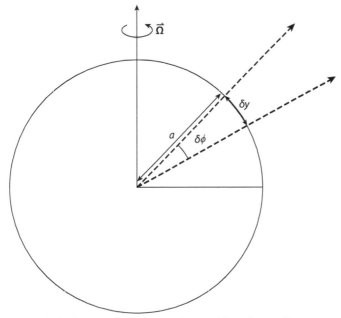

Figure B.3. Converting between the meridional coordinate, y, in local Cartesian coordinates and latitude ϕ.

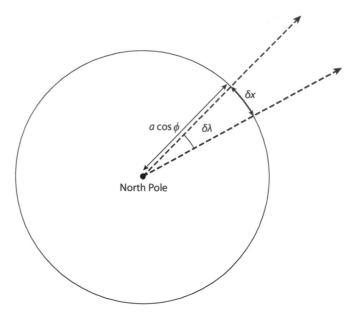

Figure B.4. Converting between the zonal coordinate, x, in local Cartesian coordinates and longitude λ.

$$\delta y = a\delta\phi \Rightarrow \frac{\partial}{\partial y} = \frac{1}{a}\frac{\partial}{\partial\phi}. \tag{B.10}$$

From Figure B.4, which is drawn from the perspective of looking down on the North Pole,

$$\delta x = a\cos\phi\delta\lambda \Rightarrow \frac{\partial}{\partial x} = \frac{1}{a\cos\phi}\frac{\partial}{\partial\lambda}. \tag{B.11}$$

The components of the horizontal velocity are

$$u = \frac{dx}{dt} = a\cos\phi\frac{d\lambda}{dt} \tag{B.12}$$

and

$$v = \frac{dy}{dt} = a\frac{d\phi}{dt}. \tag{B.13}$$

LAGRANGIAN AND EULERIAN DERIVATIVES

The application of the laws of physics, such as Newton's laws of motion applied to fluid, can be visualized by considering a fluid element, for example, a unit mass or volume of air of ocean water known as a *parcel*. Two perspectives can be taken.

The *Lagrangian perspective* considers the parcel as a physical entity to be followed in space, exactly as one would approach the classic physics problem of a block on an inclined plane.

The velocity of a parcel in the x-direction (zonal direction, or east–west direction) in local Cartesian coordinates is the Lagrangian derivative dx/dt because it tracks the changing location of the parcel with time. It is also called the "material derivative" or the "substantial derivative." The net balance of forces, or the net heating, of the parcel is calculated according to the governing equations, and the resulting change in velocity or temperature with time is expressed by the Lagrangian derivative d/dt.

Changes in any dependent variable, such as temperature, water vapor, density, or salinity, can be expressed in the same way. For example, dT/dt is the change in temperature "following the parcel." The Lagrangian perspective is satisfying in a way, partly because it is so similar to the approach taken in introductory physics texts. But it can be awkward in applications to a fluid, because the positions of many parcels must be tracked and, in the real world, the parcels stretch and distort and mix with the surrounding fluid over time. More convenient, and a better match with observing systems, is the Eulerian perspective.

The *Eulerian perspective* considers time rates of change within a fixed coordinate system. In this case, the temperature at a given location can change due to a net heating at that specific location but also because warmer or cooler air or water is transported by the circulation to that location. The local time rate of change—the Eulerian time rate of change—is denoted by the partial derivate $\partial/\partial t$.

To translate between the Lagrangian and Eulerian derivatives, consider the definition of the total differential, here applied to a temperature field that varies in space and time, $T(x,y,z,t)$:

$$dT \equiv \frac{\partial T}{\partial x}dx + \frac{\partial T}{\partial y}dy + \frac{\partial T}{\partial z}dz + \frac{\partial T}{\partial t}dt. \qquad \text{(C.1)}$$

Dividing through by the time interval dt we obtain

$$\frac{dT}{dt} \equiv \frac{\partial T}{\partial x}\frac{dx}{dt} + \frac{\partial T}{\partial y}\frac{dy}{dt} + \frac{\partial T}{\partial z}\frac{dz}{dt} + \frac{\partial T}{\partial t}\frac{dt}{dt}$$

$$= u\frac{\partial T}{\partial x} + v\frac{\partial T}{\partial y} + w\frac{\partial T}{\partial z} + \frac{\partial T}{\partial t}, \tag{C.2}$$

or

$$\frac{\partial T}{\partial t} = \frac{dT}{dt} - \vec{v} \cdot \nabla T. \tag{C.3}$$

The last term in Eq. C.3 is called the advection term. It represents the influence of the transport of heat to a location by the circulation. (See exercise 6.7.)

INDEX

Absorption, 70–76, 154–160; bands, 75; spectra, 75–76
Absorption coefficient (k), 83
Absorptivity (a), 67, 75
Adiabatic lapse rate (Γ), 118
Aerosols, 64, 153, 160; indirect effects on climate, 170; radiative forcing, 162–163
Age of ocean water, 142
Agulhas Current, 29–30
Albedo (α), 69, 85, 153; clouds, 85; dependence on solar zenith angle, 10–102; water, 100–102
Amazon Basin, precipitation, 52
Amazon River, 28
Ammonia volatilization, 184
Angular momentum, 110
Antarctica: bottom water formation, 32–33, 142; CO_2 measurements, 155–156; ice cores, 156; ice sheets, 42–43, 179; sea ice, 29; surface heat balance, 106; temperature, 11–13; topography, 5–8
Antarctic circumpolar current, 30–31
Anticyclonic, 21
Arctic Ocean: currents, 30, 141; glaciers, 43; salinity, 28–29, 89–90, 141; sea ice, 44–46; surface heat balance, 106; temperature, 11–13
Arctic Oscillation (AO), 61–62
Atlantic Ocean: air temperature, 11; currents, 29–31, 140; deep water formation, 31–33,
141–142; oxygen saturation, 142–143; precipitation, 36; salinity, 28–31; sea surface temperature, 23–25, 87; vertical distribution of temperature, 26–27. *See also* Thermohaline circulation
Atmospheric windows, 75

Benguela Current, 29–31
Biomass burning, 156, 158
Biosphere, 2; role in atmosphere's evolution, 46–47
Blackbody radiation, 66–71
Boundary currents, 31; intensification, 140
Bowen ratio, 100
Brazil Current, 29–30
Brightness temperature, 87
Brine exclusion, 142

Caballing, 145–146
California Current, 29–31
Canary Current, 29–31, 140
Carbon cycle, 182–183
Carbon dioxide (CO_2), 73–76, 153–156; absorption spectrum, 77; anthropogenic sources, 155–156; carbon cycle, 182–183; greenhouse gas, 75; ice age, 156; measurements, 154, 162; radiative forcing, 162; residence time, 154, 162; shortwave absorption, 73
Cartesian coordinates, 109, 191–193

Milton Keynes UK
Ingram Content Group UK Ltd.
UKHW050258210824
447067UK00001BA/4